Ethics and Integrity in Educational Contexts

Volume 10

The aim of this series is to provide an authoritative series of books on topics relating to ethics and integrity in educational contexts. Its scope includes ethics and integrity, defined in broad and inclusive terms, in educational contexts. It focuses on higher education, but also welcomes contributions that address ethics and integrity in primary and secondary education, non-formal educational contexts, professional education, etc. We welcome books that address traditional academic integrity topics such as plagiarism, exam cheating, and collusion.

In addition, we are particularly interested in topics that extend beyond questions of student conduct, such as

- Quality assurance in education;
- Research ethics and integrity;
- Admissions fraud;
- Fake and fraudulent credentials;
- Publication ethics;
- Educational technology ethics (e.g., surveillance tech, machine learning, and artificial intelligence, as they are used in education);
- Biomedical ethics in educational contexts;
- Ethics in varsity and school sports.

This series extends beyond traditional and narrow concepts of academic integrity to broader interpretations of applied ethics in education, including corruption and ethical questions relating to instruction, assessment, and educational leadership. It also seeks to promote social justice, diversity, equity, and inclusion.

The series provides a forum to address emerging, urgent, and even provocative topics related to ethics and integrity at all levels of education, from a variety of disciplinary and geographical perspectives.

Mary Davis • Cláudia Baptista
Editors

Ethics and Integrity in Education (Practice)

Derived from the 9th European Conference on Ethics and Integrity in Academia

Springer

Editors
Mary Davis
Arts Humanities and Social Sciences
Oxford Brookes University
Oxford, UK

Cláudia Baptista
Instituto de Ciências Biomédicas Abel
Salazar (ICBAS), Veterinary Clinics
University of Porto
Porto, Portugal

European Conference on Ethics and Integrity in Academia
eceia, eceia2023, ECEIA, Derby, UK
2023-7-12 - 2023-7-14
https://academicintegrity.eu/conference/european-conference-on-ethics-and
integrity-in-academia-2023/

ISSN 2731-779X ISSN 2731-7803 (electronic)
Ethics and Integrity in Educational Contexts
ISBN 978-3-031-95685-0 ISBN 978-3-031-95686-7 (eBook)
https://doi.org/10.1007/978-3-031-95686-7

This Springer imprint is published by the registered company Springer Nature Switzerland AG
The registered company address is: Gewerbestrasse 11, 6330 Cham, Switzerland

Preface

It is a huge pleasure to introduce the chapters that constitute this Practice book derived from sessions at the European Conference on Ethics and Integrity in Academia in Derby, July 2023. The fourteen chapters in this collection all focus on current practices in academic integrity and offer readers insights and recommendations to apply to their own practices, including a call to action to encourage readers to engage and promote academic integrity. The chapters present a diversity of perspectives in terms of author roles (faculty, academic/research integrity lead, academic conduct officer, student support staff, independent researcher, student), organisational contexts (research offices, student support teams, academic conduct teams, national and international networks, high school), geographical contexts (Australia, Canada, India, Ireland, Romania, Sweden, Türkiye, UAE, UK, Ukraine) and academic integrity practices (artificial intelligence tools, communities, faculty attitudes, inclusion, networks, publishing, research offices, student collaboration, academic conduct breaches). At the same time, the chapters are unified as a collection of effective practices which present the activities of those in academic integrity roles to uphold, promote and support academic integrity.

The book is organised under 4 main themes of academic integrity practice: innovation and collaboration, community and culture, stakeholders in academic integrity and awareness of breaches and ethics. This thematic organisation is presented in the Fig. 1 Infographic, which illustrates both the book themes and the sub-themes which interconnect across chapters.

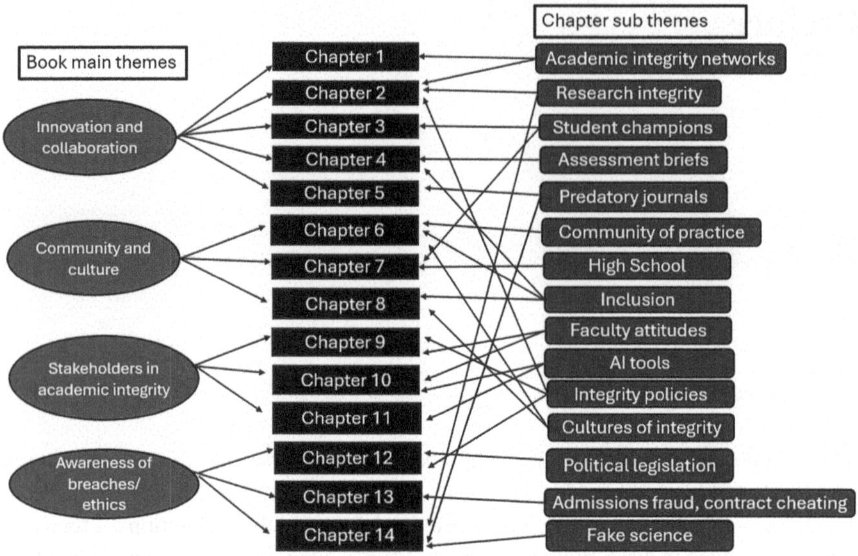

Fig. 1 Infographic of book themes and chapter sub-themes

The book themes and chapter sub-themes are introduced below.

The first five chapters discuss **innovations in practice through collaborating** to set up new initiatives, structures, roles and resources. Lancaster, Allen and Davis discuss the setting up of an academic integrity network of colleagues from different institutions in a geographically similar area in the UK, focusing on the benefits and opportunities to discuss academic integrity issues that members are facing, share advice and practice and work together on projects. Bhattacharyya, Mahapatra, Palakodeti, Ladher, Sane and Ramanathan present the process of setting up a multi-institutional research ethics and integrity office in India, highlighting the significant challenges in terms of implementing research integrity policies and the limited resources for engaging junior researchers who lack knowledge about raising scientific misconduct concerns. Khan and Hemnani describe the integration of a Student Academic Integrity Champion model into institutional structures in the UAE and detail the many activities students in these roles can be involved in, such as participating in conferences and workshops, discussions on policy, contributing to research and mentoring other students. Further innovations relate to resources that authors develop collaboratively: Sivasubramaniam, Razi, Khan and Kouroutsaltis review multiple perspectives from the literature to consider the possible correlation between unclear assessment briefs and student misconduct, and issues with timed assessment especially affecting the performance of students with Specific Learning Difficulties (SpLDs). They propose a set of measures to minimise ambiguous assessment briefs through consulting with colleagues, personalising rubrics, taking into account student needs, pilot testing, internal and external moderation and supportive feed-forward comments for students. In the last chapter on the collaborative innovation theme, Glendinning, Sivasubramaniam, Hamza, Miri, Bjelobaba and Verissimo, as members of the Ethical Publishing and Dissemination (EPAD) Working Group from the European Network for Academic Integrity (ENAI), document the steps they took to develop, evaluate, revise and share a checklist to help researchers to identify possible predatory journals.

Following on from collaboration, the next 3 chapters relate to a very popular similar theme among the chapters, **community building and creating cultures of academic integrity** through holistic, inclusive approaches. Casey discusses an Irish institution's Community of Practice approach to engage faculty, staff and all students in academic integrity decisions to increase student agency, provide meaningful collaboration opportunities and facilitate ongoing review. Casey finds that a further advantage of Community of Practice approaches is that vulnerable students, defined as any student with additional needs, can be better supported and gain confidence in following academic integrity. The chapter by Barnhardt, Torczon, Robertson, Sharma-Ferguson, Vijay and Thomas highlights the importance of 'peer power' in building cultures of academic integrity in high school education through research camps in India. Peer power stems from large and small group conversations between teacher and student peers, and through student mentoring where students with prior research camp experience guide less experienced students in ethical engagement in research projects to create a cycle of positive influence across the institution. Barnhardt and his colleagues argue that the influence on peers created through the research camps can make a crucial difference to maintaining academic integrity in the era of GenAI. My own chapter (Davis) aims to demonstrate the need to include all students and all staff in academic integrity policies and practice, especially through the use of Universal Design for Learning (UDL) to prioritise inclusion as a critical and achievable goal. The chapter incorporates a reflection on including everyone in practice from my keynote presentation at the conference in Derby, in which I involved the entire audience in discussion of their academic integrity journeys, impacts, scholarship and development, as a means of demonstrating the practice of inclusion in academic integrity.

The theme of the subsequent 3 chapters moves to gathering the **views and perspectives of stakeholders in academic integrity** (students, educators, other staff). In two Canadian institutions, Scholle-Cotton and Childs discuss faculty attitudes, beliefs and actions related to academic conduct. They found that the majority of faculty prioritised supporting students and agreed about the importance of upholding academic integrity, but differed in terms of willingness or reluctance to use formal channels to report academic conduct cases. The factors that influence their actions include prior administrative roles and familiarity with policies. Using a UK university case study, De examines student and staff experiences and viewpoints on academic integrity and current use of artificial intelligence (AI). De foregrounds the view of staff that diverse educational and cultural backgrounds strongly impact student understanding of academic integrity and highlights the critical role of learning development staff, especially in supporting international postgraduate students and fostering ethical use of AI. In another UK study, Glendinning reports on the insights from a wide-ranging consultation on student use of AI tools across their institution over a 6-month period, mainly through 17 focus groups, predominantly involving staff, but also some student representation. The consultation assisted in the development of a categorisation model for different AI tools and the guidance to staff and students about the appropriate use of AI to support learning.

The last chapters are based on the final theme of **shedding light on and investigating breaches** of academic integrity and ethics. Popoveniuc draws attention to the political interventions in plagiarism in Romania to protect personal, professional and institutional reputations. Popoveniuc spotlights legislative plans to exempt some political figures from having their doctoral theses examined for plagiarism or from experiencing any consequences, and explains that this situation has been created by factors unique to Romania in a weak education system, a concentration of top-down power in society and a lack of belonging and responsibility at university level. Turning to student breaches, Dahl and Ng reveal their in-depth investigation into a link between admissions fraud and contract cheating operations in an Australian institution. They highlight evidence of fraudulent transcripts, predatory behaviour by education agents and sub-agents, and high fees paid by students for the promise of admission to the university. Thereafter, the authors demonstrate an example of a student who submitted both fraudulent admissions documentation and assignments that contained evidence of using cheating services, such as their learning management account being accessed simultaneously from multiple international locations. In the final, highly novel chapter, Artyukhov directs the reader to reflect on experiences of fake science through presenting a set of episodes in Ukraine: an experience of a predatory conference, a 'pointless' article publication in a predatory journal, a portrait of those who operate predatory conferences and journals and finally an imaginary dialogue between a scientist and fake science.

We hope readers of this Practice book will be inspired by the calls to action and recommendations of this volume (in order of chapters) to:

- Establish regional networks of academic integrity and sustain them through regular online meetings to discuss relevant current topics for members and provide space for sharing and collaborating (Lancaster, Allen and Davis)
- Develop multi-institute research integrity offices through coordinating with many established global networks of research integrity and focusing on training researchers in integrity (Bhattacharyya, Mahapatra, Palakodeti, Ladher, Sane and Ramanathan)
- Integrate an institutional model of Student Academic Integrity Champions to engage students actively in debates, events, policies, scholarship and mentoring (Khan and Hemnani)
- Ensure assessment briefs have clear instructions regarding consistent, fair grading criteria aligned with learning outcomes, that students have access to exemplars to understand expectations and are supported appropriately with assessment and that assessment briefs are regularly reviewed, so that they are fit for purpose (Sivasubramaniam, Razi, Khan and Kouroutsaltis)
- Raise awareness of predatory journals by evaluating them through a checklist to examine journal titles, finance, associations, quality and scope, indexing,

publishing practices, communication and copyright (Glendinning, Sivasubramaniam, Hamza, Miri, Bjelobaba and Verissimo)
- Create a Community of Practice approach to academic integrity within institutions to engage all faculty, staff and students to foster new innovations, adapt to change and support students who may be considered vulnerable (Casey)
- Design research camps to build and sustain cultures of integrity through peer influence, in which students with prior experience mentor new students (Barnhardt, Torczon, Robertson, Sharma-Ferguson, Vijay and Thomas)
- Prioritise inclusion in academic integrity by making it relevant, accessible, engaging and central to all students and staff, employing UDL, incorporating academic integrity into every programme accessibly and developing inclusive approaches to student practice with AI (Davis)
- Engage faculty in academic integrity procedures and policies by providing clear and complete information, emphasising the importance of these procedures and policies and inviting their feedback (Scholle-Cotton and Childs)
- Collaborate with and empower Learning Development colleagues to support staff and students with understanding and following academic integrity and managing concerns with AI (De)
- Encourage institutional senior management to provide a clear statement on use of AI to guide students and staff, and liaise with professional bodies and industry contacts about their views of AI requirements, to help prepare students for work beyond academia (Glendinning)
- Strengthen institutional responsibility for plagiarism and exchange good practice with other countries, through networks, events and campaigns to combat plagiarism (Popoveniuc)
- Examine the potential link between admissions fraud and contract cheating by liaising and sharing information between the relevant institutional teams, as well as exchanging between institutions to combat the continuation of fraudulent admissions (Dahl and Ng)
- Approach scientific information with a critical eye, verify sources, stay informed, encourage others and expose fake science by reporting it to relevant organisations (Artyukhov)

As can be seen by the above overview and calls to action, this volume of chapters about academic integrity practice is highly varied, with nine different national contexts presented and diverse educational environments from high school to research centres with a range of key themes in terms of collaboration, community, stakeholder views and examination of breaches. Thus, the book aims to raise awareness of diverse and important issues in academic integrity practice contexts, provide practical recommendations and engage all readers through calls for action to improve and promote academic integrity.

Oxford, UK — Mary Davis
Porto, Portugal — Cláudia Baptista

Acknowledgements

As the lead editor of the research and practice volumes of this book, I would like to express my heartfelt gratitude to all those who contributed to its completion. This book is the culmination of the engaging discussions, presentations and workshops that took place during the 9th European Conference on Ethics and Integrity in Academia (ECEIA-2023), which I had the privilege of hosting at the University of Derby, UK, in 2023. This project would not have been possible without the collaborative efforts of the authors, reviewers and practitioners who shared their invaluable insights, research findings and expertise. Organised under the auspices of the European Network for Academic Integrity (ENAI), this conference highlighted critical issues at the intersection of emerging technologies and academic integrity.

The editorial team extends heartfelt gratitude to Dr Tomáš Foltýnek, President of ENAI, and its Board members, as well as to Prof. Kathryn Mitchell, Vice-Chancellor of the University of Derby, and Prof. Chris Bussell, Dean of the School of Science and Engineering, for their invaluable support in making this conference and subsequent publication possible. The idea of transforming the conference proceedings into a book emerged from the importance of the themes explored during the event, particularly the impact of emerging technologies on academic integrity. When presented to Springer, the concept was met with enthusiastic encouragement, for which we are sincerely grateful to the editor Astrid Noordermeer. This support enabled us to approach contributors with a vision for the book, inviting them to expand their insights and arguments beyond their initial presentations. The response was overwhelmingly positive, leading us to produce not one but two volumes: one focused on research and the other on practical applications in academic integrity and ethics.

We express our deepest appreciation to the authors whose contributions form these volumes. Their commitment to scholarly excellence has enriched this work immeasurably. We also thank the international experts and peer mentors who reviewed the chapters, offering constructive feedback and guidance that significantly enhanced the quality of submissions.

As lead editor, I was fortunate to collaborate with an exceptional editorial team: Prof. Sarah Elaine Eaton, Dr Irene Glendinning, Prof. Mary Davis, Dr Claudia Baptista and Dr Beatriz Moya, whose expertise and dedication were instrumental in shaping this book. – I was merely ‘standing on the shoulders of giants’. I extend my deepest gratitude to Dr Vadivel Parthsarathy for his invaluable support during the final stages of this project, helping me with the submission of the Research volume. His attention to detail and dedication during these last few days were essential to achieving the highest standards in the submission process.

This project embodies the spirit of scholarly collaboration across borders, featuring contributions from authors worldwide. Their diverse perspectives and expertise are reflected in the 24 chapters of both volumes, spanning critical areas of ethics and academic integrity. We hope that readers will find these volumes insightful and thought-provoking, furthering the global conversation on these vital topics.

Dr. Shiva Das Sivasubramaniam
(On Behalf of the Editorial Team)

Contents

Contributors

Steph Allen Bournemouth University, Poole, UK

Artem Artyukhov Sumy State University, Sumy, Ukraine
University of Economics in Bratislava, Bratislava, Slovakia

Bradford Barnhardt Woodstock School, Mussoorie, India

Sabuj Bhattacharyya Institute for Stem Cell Science and Regenerative Medicine, Bengaluru, India

Sonja Bjelobaba Uppsala University, Uppsala, Sweden

Elva Casey Maynooth University, Maynooth, Ireland

Ruth A. Childs University of Toronto, Toronto, ON, Canada

Darcey Dahl University of New South Wales, Sydney, NSW, Australia

Mary Davis Arts Humanities and Social Sciences, Oxford Brookes University, Oxford, UK

Debbie De Aston University, Birmingham, UK

Irene Glendinning Independent Researcher, Leicestershire, UK

Muaawia Ahmed Hamza King Fahad Medical City, Riyadh, Saudi Arabia

Neha Hemnani University of Wollongong in Dubai, Dubai, UAE

Zeenath Reza Khan University of Wollongong in Dubai, Dubai, UAE

Dimitrios Kouroutsaltis University of Dundee, Dundee, UK

Raj K. Ladher National Centre for Biological Sciences, Bengaluru, India

Thomas Lancaster Imperial College London, London, UK

Biswa Bhusana Mahapatra National Centre for Biological Sciences, Bengaluru, India

Siavash Miri Brieflands, Heerlen, Netherlands

Brandon Ng University of New South Wales, Sydney, NSW, Australia

Dasaradhi Palakodeti Institute for Stem Cell Science and Regenerative Medicine, Bengaluru, India

Anusuya Panchagatti Woodstock School, Mussoorie, India

Bogdan Popoveniuc University Stefan cel Mare of Suceava, Suceava, Romania

Arvind Ramanathan Institute for Stem Cell Science and Regenerative Medicine, Bengaluru, India

Salim Razi Çanakkale Onsekiz Mart University, Çanakkale, Türkiye

John Robertson Hanoi International School, Hanoi, Vietnam

Sanjay P. Sane National Centre for Biological Sciences, Bengaluru, India

Lydia Scholle-Cotton Queen's University, Kingston, ON, Canada

Vandana Sharma-Ferguson Woodstock School, Mussoorie, India

Shiva Das Sivasubramaniam University of Roehampton, London, UK

Rahima Thomas Woodstock School, Mussoorie, India

Asha Torczon Cambridge University, Cambridgeshire, UK

Ana Cristina Veríssimo University of Porto, Porto, Portugal

Chapter 1
The Benefits of Academic Integrity Networks: Exploring the London and South East Academic Integrity Network

Thomas Lancaster, Steph Allen, and Mary Davis

Abstract Academic integrity networks are of value to everyone in the educational community, providing space for members to exchange ideas, discuss problems, and to be informed about changes within the academic integrity ecosystem. This chapter discusses the benefits of academic integrity networks, regarding one particular network, the London and South East Academic Integrity Network (LSEAIN). The chapter explores how LSEAIN was established in May 2021, how the network operates, and the strategies used to develop and promote the network, for which meetings operate under the Chatham House rule.

The chapter also includes case studies of activities that LSEAIN members have been involved with, many of which would not have been possible without having the network as a meeting point. It considers examples involving the heavy involvement of members, including the authors themselves, with the International Day of Action for Academic Integrity and how members joined together to gain funding to work on improving inclusive academic integrity practice. Members benefit from the cross-discussions of both academic and professional services staff. They get to meet people from different disciplines and institutions who they might not otherwise have met. The regular LSEAIN meetings have allowed participants to consider emerging academic integrity issues, such as the use of online proctoring software during the COVID-19 years, and more recently, opportunities to respond with integ-

T. Lancaster (✉)
Imperial College London, London, UK
e-mail: t.lancaster@imperial.ac.uk

S. Allen
Bournemouth University, Poole, UK
e-mail: stephaniea@bournemouth.ac.uk

M. Davis
Arts Humanities and Social Sciences, Oxford Brookes University, Oxford, UK
e-mail: marydavis@brookes.ac.uk

M. Davis, C. Baptista (eds.), *Ethics and Integrity in Education (Practice)*, Ethics and Integrity in Educational Contexts 10,
https://doi.org/10.1007/978-3-031-95686-7_1

rity to the widespread availability of generative AI. The chapter is intended to be useful to others looking to establish their own academic integrity networks.

Keywords Academic integrity · Networking · Members · Opportunities

Introduction

This chapter provides a case study of the London and South East Academic Integrity Network (LSEAIN), an example of an academic integrity network designed to allow members to share ideas and experiences, develop collaborations, and to grow the wider academic integrity community. LSEAIN is not the only academic integrity network of this type, but the information published about the benefits of academic integrity networks and how these can be established and managed is limited. Other similar documented examples cover the formation of the Academic Integrity Ontario Group in 2008, which later became known as the Academic Integrity Council of Ontario (McKenzie et al., 2020), as well as the development of the Manitoba Academic Integrity Network in 2019 (Stoesz et al., 2020).

This account of the approach taken by LSEAIN is intended to provide inspiration and ideas for the formation of other such networks. These can take place locally at universities, at discipline level, be geographically located, and come with the option to involve students. The chapter is based on a presentation given at the European Conference on Ethics and Integrity in Academia 2023 (Lancaster et al., 2023).

LSEAIN is an informal and unfunded network, which was established in May 2021. The network is largely aimed at people working in higher education within the geographical region of London and South East England, although geographical boundaries are not strictly enforced. Typical members include teachers and lecturers, learning developers, student conduct and appeal team members, academic integrity leads, and academic integrity researchers, although LSEAIN is not strictly a research network. Details of how to join LSEAIN are available on the website (LSEAIN, n.d.). Other geographical academic integrity networks operate within the United Kingdom, although there are no formal links between them (Networks, n.d.).

The chapter takes the following format: first, the wider benefits of developing networks within an educational setting are discussed; second, the formation and development of LSEAIN is detailed, along with the structure of LSEAIN meetings and examples of the topics discussed; third, successes that have come about as a result of LSEAIN collaborations are provided; finally, the chapter provides ideas and inspiration for others looking to develop and grow their own academic integrity network.

The Benefits of Networks

The topic of networking appears to be one of those subjects that has much written about it but somehow falls into the business self-help and motivation category sections of the library. Often identified as a "*How to* ...", "*A Guide for*..." or "*The Art of*...", an overwhelming melee of instructional, informative, spiritual, science-based, aptitude, or mastery narratives have been produced, positioning the reader as a mentee, student, learner, or apprentice in some kind of supernatural craft where mastery can only be achieved through poring over and digesting published words of wisdom. Indeed, much has been written about networking over the century starting with Simmel's (1903, cited by Gross, 2000) publication which promotes the notion that relationship-based networks are fluid and organic shaped by the interactivity of the individuals and communities (including the nonhuman) within those groups.

Whilst powerhouse individuals proclaim the advantages of mastering networking techniques to 'win friends and influence people' (Carnegie, 1936) or 'get your back '(Ferrazzi, 2009), networking has been demonstrated through the lens of business, with a view to money making prospects (Misner & Hilliard, 2017) whereas others (Ottewill et al., 2005) focus on a scholarly and civil approach: pedagogic innovation, knowledge, advice, and support without expecting a return. Yet it is by working together, as Ferrazzi (2005) highlights, that small interactions can lead to bigger things and broader change, such as a conversation about an emerging project that can lead to a conference paper or keynote speech. It is essential to be proactive, such as sending an enquiring email. Whilst peers for specialist interests can be found—the online bookstores and social media have an ocean of offerings—little evidence has emerged since Stoesz et al. (2020) on the benefits of networking in the academic integrity community. Therefore, this chapter aims to contribute insights and advice on academic integrity networks.

It could be argued that clarifying the benefits of networking and aligning to academic integrity is essential to avoid misunderstandings between benefits and values. Table 1.1 below presents the key elements and benefits of networking, drawn from an overview of the self-help literature such as Cain (2013), Carnegie (1936), Ferrazzi (2005, 2009), Garner (2019), Gladwell (2000), Grant (2013), Hoffman and Casnocha (2012), Misner and Hilliard (2017), and aligning to the International Center for Academic Integrity (ICAI, 2014) Six Fundamental Values that underpin Academic Integrity. This examination of self-help and academic integrity values visualises how the two are interconnected and how networking can be aligned with academic integrity.

As someone once said "*It's not what you know, it's who you know*" and in the academic integrity community, all members connect with others recognising integrity as a binding underpinning force that maintains their standing in the sphere.

In a digital landscape which is rapidly reframing connectivity and inter-relationships, networking in the human and social sense has never been more deep-felt, desired, or dynamic. In the absence of a specific network, opportunities can be created to develop one to address the needs of an existing or potential community.

Table 1.1 Elements and benefits of networking in terms of the ICAI (2014) Fundamental Values

Key Elements of Networking	Key Benefits of Networking	Networking with integrity: shared values. Practitioner of the ICAI Six Values:
Values: Being of service expecting nothing in return, spirit of generosity	Opportunities, recommendations, and referrals as an attractive networking contact for upcoming opportunities such as collaborative writing, co-editing, guest speaker.	ICAI values: Honesty, trust, fairness, respect, responsibility, and courage.
Relationship building and genuine connections: Mutually beneficial	Collaborative partnerships and supportive network. Introductions to groups and colleagues.	ICAI value: Trust
Widening contacts/ collaborators/partners	Community of peers, new opportunities from introductions, responding to emails, reaching out, and following through with the expectations and tasks.	ICAI values: Responsibility and honesty.
Information exchange	Knowledge sharing, problem-solving, sector trends. Shared research findings, referencing sources, plagiarism avoidance.	ICAI values: Responsibility and honesty.
Consistent ongoing positive interactions	Professional development, and increased visibility. Keeping in touch, attending meetings, being present, truthful, and respectful, adhering to policies.	ICAI values: Respect and responsibility.
Authenticity; demonstrating interest	Listening to feedback, having empathy, open communication, knowledge sharing, accepting diversity, differences and recommendations. Critical reviewer for others working in the field, maintaining confidentiality.	ICAI values: Respect and trust.
Growing others	Advice and guidance, career advancement. Upholding ethical principles through model behaviours.	ICAI values: Honesty, trust, fairness, respect, responsibility, courage.

To get the best from a network, finding the best approach may evolve from trial-and-error experiences or recommendations.

Network Formation and Operation

The original plans for LSEAIN came about during the pandemic period when it was noted that there was no central outlet for interested parties around the London and South East England region to meet and discuss academic integrity. An initial request for interest was circulated using the JISCMail Plagiarism List in April 2021 (Lancaster, 2021) and through social media channels. An initial meeting was arranged for May 2021. The first meeting and all subsequent meetings have taken place using Zoom.

The founding members collectively determined how the network would operate. It was agreed that members would not be formally representing their academic institution and that multiple members could join from the same institution. Although the network was founded to serve London and South East England, guest members from other parts of the UK have also been allowed to informally join meetings and communications.

Typically, meetings take place around five times per year, all during the main academic year. Email discussions take place between scheduled meetings. Details of members, meeting notes, and other resources, are held on a private shared Google Drive. Meetings are conducted under the Chatham House Rule (n.d.), officially defined as "*when a meeting, or part thereof, is held under the Chatham House Rule, participants are free to use the information received, but neither the identity nor the affiliation of the speaker(s), nor that of any other participant, may be revealed.*"

A typical meeting is scheduled for 2 h, with the following structure:

- Welcome to new members
- Member updates
- Working group updates
- Discussion of a main topic agreed upon by the membership
- Open discussion

Meetings are discussion-based and do not include formal presentations.

Each meeting includes the discussion of a main topic determined by the members, usually at the previous meeting. Due to the timeframe for which LSEAIN has been in operation, many main topics have fallen within the wider themes of COVID-19 and generative artificial intelligence. Examples of topics discussed at previous meetings include:

- Online proctoring and online assessment
- Training staff around academic integrity matters
- Engaging students in the academic integrity discussion
- Policy development surrounding whistleblowing and attempts to extort students
- Understanding discipline differences in academic integrity approaches and providing support for non-textual disciplines
- Designing assessments to help students avoid academic integrity pitfalls
- Essay bots, artificial intelligence and automated assessment-producing systems
- Developing understandable, inclusive, accessible, and future-proof policies
- Incorporating the student voice into the agenda
- Professional values and employability as these relate to academic integrity
- Impact of Contract Cheating Legislation in England
- Generative AI detection versus human detection
- Academic integrity and reputation management

Many discussions have led to recommendations shared within LSEAIN, which individual members have been able to take back to their institutions whilst respecting the Chatham House Rule. Although recommendations can soon become outdated and it is not possible to directly share these within this chapter, one important result of network discussions is sharing community insights. After discussion, members can take insights back to their institutions and report on other universities that have identified good practices, which can be a powerful driver for change across the sector.

Network Successes

LSEAIN as a whole has not made any bids for external funding, but it has spawned several working groups and members have developed their offshoots from LSEAIN activity. This section shares four selected LSEAIN success stories.

Academic Integrity Speaker Series

Picture this.

A warm summer's day during the COVID-19 pandemic where physical isolation was a key tenet of staying 'safe' and where communication was transported to online modes for connectivity. Without repeating all that has gone before about the rise in the ennui of being lost and confused, the endless unknowing when things would return to normal, and even the concept of what 'normal' even was or would develop into raised several deep and reflective questions in the HE sector: if students are studying, how are they studying, and what do they understand about academic integrity in an unprecedented time of geographical and physical separation. Additionally, how were educators coping and addressing new approaches to what could be perceived as student academic misconduct behaviours when human connectivity was disjointed, dissonance set in, and detachment from teaching and learning for many, was noticeable. For some, particularly educators viewing emerging trends in potential offences, wanting to assess whether early-stage misconduct was an isolated event within institutions or compare and assess whether a rising threat to integrity was impending, the need to connect with other academics and professionals was critical. In May 2021, one of the authors (Allen) heard about LSEAIN from some colleagues who had an academic integrity network in another part of the country and was signposted to contact Thomas Lancaster and LSEAIN. Whilst out of the geolocation of the group being welcomed in was evidence of a culture of openness and sharing. Attending several meetings helped establish the strengths of the group and the weaknesses of institutional activities at a time of perplexing uncertainty. It was also the place where co-authors met (Allen and Davis), amongst others, which would become important in time. The ongoing discussions evidenced that our group was just one satellite of communities asking the same questions.

At the same time, a connection with a community of educators in Canada was also discussing similar topics and after a few emails and joining their online sessions, it was clear that something similar in the UK was required and so was borne the fledgling Academic Integrity Speaker Series: Tales of the Unexpected (AISS). The key idea was to gather narratives from educators who were exploring and publishing through the pandemic on experiences, research, and technical aspects of academic misconduct during the pandemic but also crystal balling future needs for educators and students as hybrid education was developing.

Networking through LSEAIN, already connected with the Canadians and other global parties, a collection of webinars evolved with emergent 'rock-star' speakers, including co-authors Lancaster and Davis, willing to offer their time, narratives, experiences, guidance, and advice to anyone who might be interested. Further, connections offered suggestions for future topics and potential speakers who viewed the misconduct landscape ahead with astute insights and clarity. Time is precious and they were willing to share it evidencing a key networking element: the spirit of generosity. At a time of new busyness, sharing knowledge was key. Networking groups exist to establish mutual respect and understanding rather than a series of transactional exchanges, and has been evident throughout amongst a community of colleagues, students, and mentors, an example of how educators work institutionally but during this transitory era, the inclusive Speaker Series became a quiet international phenomenon attracting audiences with similar interests seeking authentic connection-building opportunities for future advantage and to create their own networking group. LSEAIN and AISS opened opportunities to connect with other high-profile networking groups globally, expanding the interest, guiding, supporting, and empowering others to share network goals; the mutually beneficial approach contributed to the greater good and continues to do so. The foundation of LSEAIN added to the number of regional academic integrity networks, and since then, further new networks have evolved in the UK and across the globe. In this way, forming networks has encouraged a robust support group of individuals in the pursuit of a new connectedness, away from enforced seclusion. It has also resulted in additional Academic Integrity Speaker Series opportunities which will evolve as long as like-minded educators, who cannot be thanked enough, continue to hold the integrity banner, share values, speak truths, and network like friends for mutual collegiate benefits.

QAA Accessibility and Inclusion Resources

As an active network, LSEAIN has facilitated significant opportunities for networking and finding colleagues with similar interests and approaches to academic integrity. One highly productive example is four network members with a similar strong interest in inclusive educational approaches to academic integrity, from Oxford Brookes University, University of Southampton, Bloomsbury Institute London and University of Westminster, who won a competitive bid for QAA funding of a Collaborative Enhancement Project. The project, led by one of the co-authors (Davis), was entitled 'Improving student learning by linking accessibility/inclusion with academic integrity'. Each of the four LSEAIN members became an institutional lead for the project, with colleagues in academic integrity working with inclusion experts, Student Union representatives and students. The whole project team involved about 25 colleagues who worked productively and enthusiastically together over 16 months through fortnightly meetings. The team collaboration led to them successfully developing 14 inclusive and accessible educational resources for teaching academic integrity to students and staff, as well as facilitating student voice videos and creating a student CHAMPION model (**C**ollect information, **H**one the definition, **A**ttract, recruit and select, **M**anage and prepare, **P**ilot the role, **I**nvolve in academic integrity, **O**bserve and reflect, **N**urture the role) to assist institutions in providing a role for students in academic integrity (QAA, 2023). The project sought to recognise and support the needs of all students, especially neurodivergent, international and widening participation students, who tend to be over-represented in academic conduct cases (Davis, 2023); thus the resources were created using Universal Design for Learning principles (CAST, 2018) to ensure they were fully accessible to all students.

To date, the project resources page has had 6000 views and 4500 downloads of resources across 43 Higher Education Institutions (HEIs). Members of the project team held an all-day international launch event with 165 participants (from 89 UK HEIs and 12 international HEIs) who rated the resources highly and commented that the event raised their awareness of the challenges some students may have with academic integrity resources. Project team members have also contributed findings and discussed the resources at five conferences, including the Advance HE Teaching and Learning Conference and the QAA Quality Conference. As a direct consequence of the success of the project, the project lead was invited to join the QAA UK Academic Integrity Advisory Group. All these impacts stemmed from the initial conversations at LSEAIN meetings!

International Day of Action for Academic Integrity

The interactions between members of LSEAIN have also been influential in their significant participation in the ICAI International Day of Action for Academic Integrity (IDoA) in the last 2 years, including one of the co-authors (Davis) becoming co-chair of the event in 2023. Discussions and updates in LSEAIN meetings to promote the IDoA in 2022 and 2023 have led to several members contributing directly to the event in dialogues and panel discussions, giving them a global platform to discuss their work and share their insights on academic integrity. In addition, many members engaged by attending sessions in the programme of live events. Interest is growing in this event among members, and this may help to encourage more academic integrity colleagues in the UK to run IDoA events within their institutions, thus growing the momentum for action on academic integrity to become a collective, institution-wide responsibility, as per the Academic Integrity Charter for UK Higher Education (QAA, 2020). The importance of the network in supporting

IDoA cannot be underestimated, as it provides a key means of promoting and engaging members, who then promote and engage others at their institutions and beyond, providing the vital grassroots for academic integrity messages to be grown and shared widely.

South & South West Academic Integrity Network

LSEAIN has shown its worth and value to its members. Like so many other groups, the membership evolves and adapts according to jobs and other factors. LSEAIN has been open and generous to all those in the group, including those out of the area. One such unrepresented area was the South & South West of England, rich with universities and colleges, but not fully connected. From an already established pattern of meetings through the academic calendar, and to address the need for such a group for a wider community, the LSEAIN inspired the creation of the 'say what it does on the tin' South & South West Academic Integrity Network (SSWAIN, n.d.). The membership comprises Learning Developers, Academics, Learning Technicians, Students, and Professional Staff within the region. Operating under the same Chatham House Rule, it runs informally, both online using Microsoft Teams or in-person, with opportunities for short drop-ins rather than scheduled events. The format varied from LSEAIN in the early days to determine preferred connection times with continued reviews for optimum meeting times. The network is expected to continue to expand and new members within the region are welcome.

Recommendations

One of the largest challenges with developing an academic integrity network is recruiting an active and engaged membership and keeping the momentum going. LSEAIN has settled into a rhythm with a core of engaged members who collaborate well together, but there always remains a need to keep promoting and marketing the network.

This chapter would be incomplete without an acknowledgement that developing a new academic integrity network is not always without challenges. During the time that LSEAIN has run, the organisers have had to make some difficult decisions regarding fairness of access and ensuring that confidentiality is maintained. Despite that, there is strong agreement that networks of this type offer immense value for everyone concerned.

This recommendations section contains a short checklist of questions that anyone thinking about launching a network should consider, along with associated considerations.

Who Is the Network Aimed at?

LSEAIN is targeted towards people working at academic institutions in London and South East England. Other networks may focus on an interest within academic integrity (or beyond). They may operate in a specific subject area, or within an institution.

Within LSEAIN, the members chose to exclude commercial representatives, such as those people working for providers of commercial academic misconduct detection solutions. Legal representatives, such as educational barristers, have also not been able to join. This can prevent sponsorship opportunities.

A tougher discussion took place regarding the involvement of students. The decision of LSEAIN has been to welcome representation from Student Union organisations. Whether individual students can join or not is something determined on an individual basis.

How Can a Network Be Launched and Members Recruited?

Advertising and marketing are always difficult. The launch of LSEAIN relied on freely available mailing lists and social media posts. Members have also mentioned LSEAIN when giving presentations. The word-of-mouth network has also been powerful.

How Can a Network Be Funded?

Without relying on commercial sponsorship, LSEAIN has run purely on a voluntary basis. Meetings have been held online, avoiding travel costs. Members have occasionally met up at existing academic integrity events.

It may be that individual institutions have funds to launch academic integrity networks. One of the authors previously launched a Contract Cheating Special Interest Group through the Higher Education Academy (HEA), which attracted a small amount of seed funding, but following changes within the HEA, that group is no longer in operation.

Other activities launched through LSEAIN contacts, such as those detailed in the earlier Network Successes section, have also attracted separate funding.

How Should Meetings Operate?

LSEAIN developed a standard format for meetings, including one main discussion topic each time. Typically, meetings stick closely to that structure, but more general discussion topics can happen at any time. During Zoom meetings, members will usually raise their hands to contribute to discussions verbally or will add comments to the text chat.

Meetings are not recorded, so meeting notes can be rather ad-hoc, as they are taken alongside the discussion. All members can review and edit the meeting notes after each meeting.

The decision was made early on that meetings would not have formal slides or presentations. Other academic integrity networks within the UK operate differently, including inviting external speakers to attend. The LSEAIN decision was that there were sufficient other opportunities for members to join educational webinars, although the opportunities for extended and more private discussion within the network can mean that there is a further value to be added there.

How Can the Momentum Be Kept?

Within LSEAIN operating primarily as a voluntary network, occasionally plans can be lost behind the other tasks that are always necessary within academia. Any network needs members who will strongly drive the community. Some LSEAIN members are active researchers or lead academic integrity activity within their institution. Some members join purely because they find the topic fascinating but have little activity of their own to contribute. It seems important that all meetings are welcoming to new members.

The format of LSEAIN meetings is not set in stone. It is very much hoped that LSEAIN will continue in some capacity even if the organisers are unable to fulfil their current leading roles.

Conclusions

This chapter serves as a demonstration of the value of academic integrity networks. LSEAIN has rapidly gained traction within the UK and beyond as a source of idea sharing, shared support and as a valued sounding board for members.

The co-authors hope this chapter motivates those who wish to develop or expand their network and will be inspired to overcome potential barriers to set up their own, local networks. They also encourage readers to join an existing academic integrity network to support future generations of educators and mentees as the eduscape evolves. Reaching out and making connections through networks help the academic integrity community to expand and thrive.

References

Cain, S. (2013). *Quiet: The power of introverts in a world that can't stop talking*. Penguin.

Carnegie, D. (1936). *How to win friends and influence people*. Simon & Schuster.

CAST. (2018). *Universal design for learning guidelines version 2.2*. http://udlguidelines.cast.org

Chatham House Rule. (n.d.). *Chatham house rule*. Chatham House. https://www.chathamhouse.org/about-us/chatham-house-rule

Davis, M. (2023). Inclusion within a holistic approach to academic integrity: Improving policy, pedagogy, and wider practice for all students. In S. E. Eaton (Ed.), *Handbook of academic integrity* (pp. 1–19). Springer. https://doi.org/10.1007/978-981-287-079-7_127-1

Ferrazzi, K. (2005). *Never eat alone*. Crown Publishing Group.

Ferrazzi, K. (2009). *Who's got your Back*. Crown Publishing Group.

Garner, J. (2019). *It's who you know: How to make networking work for you*. Wiley.

Gladwell, M. (2000). *The tipping point*. Little, Brown and Company.

Grant, A. (2013). *Give and take*. Penguin Books.

Gross, M. (2000). Classical sociology and the restoration of nature: The relevance of Émile Durkheim and Georg Simmel. *Organization & Environment, 13*(3), 277–291. https://doi.org/10.1177/1086026600133001

Hoffman, R., & Casnocha, B. (2012). *The startup of you: Adapt, take risks, grow your network, and transform your career.* Crown Publishing Group.

ICAI. (2014). *The fundamental values of academic integrity.* https://academicintegrity.org/images/pdfs/20019_ICAI-Fundamental-Values_R12.pdf

Lancaster, T. (2021, April 19). *London and SE England academic integrity network.* JISCMail Plagiarism Archives. https://www.jiscmail.ac.uk/cgi-bin/wa-jisc.exe?A2=ind2104&L=PLAGIARISM&O=D&P=2262

Lancaster, T., Allen, S., & Davis, M. (2023, July 12-14). The benefits of academic integrity networks. *Exploring the London and south east academic integrity network* [Conference presentation]. European conference on ethics and integrity in academia 2023.

LSEAIN. (n.d.). *London and south east academic integrity network.* Academic Integrity UK. https://academicintegrity.uk/networks/lseain

McKenzie, A., Miron, J., & Ridgley, A. (2020). Building a regional academic integrity network: Profiling the growth and action of the academic integrity council of Ontario. *Canadian Perspectives on Academic Integrity, 3*(1), 25–38. https://doi.org/10.11575/cpai.v4i1.72913

Misner, I., & Hilliard, B. (2017). *Networking like a pro: Turning contacts into connections.* Entrepreneur Press.

Networks. (n.d.). *Academic integrity networks.* Academic Integrity UK. https://academicintegrity.uk/networks

Ottewill, R., Riddy, P., & Fill, K. (2005). International networks in higher education: Realising their potential? *On the Horizon., 13*(3), 138–147. https://doi.org/10.1108/10748120510618169

QAA. (2020). *Academic integrity charter for UK higher education.* https://www.qaa.ac.uk/membership/membership-areas-of-work/academic-integrity/charter

QAA. (2023). *Improving student learning by joining up inclusion/accessibility and academic integrity.* Collaborative Enhancement Project–Oxford Brookes University, University of Southampton, Bloomsbury Institute and University of Westminster. https://www.qaa.ac.uk//en/membership/collaborative-enhancement-projects/academic-integrity/the-improvement-of-student-learning-by-linking-inclusion-accessibility-and-academic-integrity

Simmel, G. (1903). Die Großstädte und das Geistesleben [The metropolis and mental life]. *Jahrbuch der Gehestiftung, 9*, 187–206.

Stoesz, B. M., Seeland, J., MacLeod, P., Vogt, L., Markovics, L., & Denham, T. (2020). Creating a collaborative network to promote cultures of academic integrity in Manitoba's post-secondary institutions. *Canadian Perspectives on Academic Integrity, 3*(1), 39–48. https://doi.org/10.11575/cpai.v3i1.69763

SSWAIN. (n.d.). *South & South West academic integrity network.* Academic Integrity UK. https://academicintegrity.uk/networks/sswain/

Chapter 2
Building a Multi-Institutional Research Ethics and Integrity Office in India: Processes and Challenges

Sabuj Bhattacharyya, Biswa Bhusana Mahapatra, Dasaradhi Palakodeti, Raj K. Ladher, Sanjay P. Sane, and Arvind Ramanathan

Abstract Academic integrity is at the core of good research practice, helping to create and sustain the public's trust in scientific research. While the challenges in academic integrity, particularly scientific misconduct are global, there are factors which indicate idiosyncratic nature in frequency and magnitudes of research misconduct across geographical space. The need to maintain high standards of research ethics and integrity has led to the formation of dedicated units or offices known as "Office of Research Integrity (ORI)" or "Research Integrity Office (RIO)" in the global north. However, they are scarce in developing countries such as India. Moreover, the specific challenges faced (e.g., academic integrity policy implementation) by Indian academics are poorly documented. Addressing this issue, two leading life science research institutes in India, the Institute for Stem Cell Science and Regenerative Medicine (iBRIC-inStem) and National Centre for Biological Sciences (NCBS-TIFR) which share a common campus and research facilities with cross-campus collaborations between faculties and students, established India's first dedicated RIO in 2022. The office is primarily working with three aims: to formulate policies and protocols for using tools for academic integrity to help uphold the academic integrity at each institute, to create a process for publication data archiving to improve the transparency and reproducibility of research conducted and finally, to conduct capacity building and training programmes for all stakeholders. Here, we aim to discuss the process of setting up a multi-institute research ethics and integrity office as a best practice example of cooperation, and optimum utilisation of limited resources, as well as various measures to improve stakeholder engagement and network.

S. Bhattacharyya (✉) · D. Palakodeti · A. Ramanathan
Institute for Stem Cell Science and Regenerative Medicine, Bengaluru, India
e-mail: sabujb@instem.res.in; dasaradhip@instem.res.in; arvind@instem.res.in

B. B. Mahapatra · R. K. Ladher · S. P. Sane
National Centre for Biological Sciences, Bengaluru, India
e-mail: biswab@ncbs.res.in; rajladher@ncbs.res.in; sane@ncbs.res.in

M. Davis, C. Baptista (eds.), *Ethics and Integrity in Education (Practice)*, Ethics and Integrity in Educational Contexts 10,
https://doi.org/10.1007/978-3-031-95686-7_2

Keywords Academic integrity · Collaboration · Global south · Multi institute · Research ethics and integrity office

Introduction

Research Misconduct: A Global Challenge

Research integrity is an essential attribute of any research and good research practice forms core standards to ensure transparency and reproducibility to create and sustain the public's trust in modern research. Scientific misconduct and questionable research practices are a global problem and have been reported in both high-income countries in the global north such as the United States, United Kingdom, and Germany, as well as from lower and middle-income countries such as China and India (Fanelli, 2009; Ana et al., 2013). Thus, increased incidents of scientific misconduct in the past few decades received considerable attention from the global research community (Smith, 2006; Gupta, 2013). A recent meta-analysis on misconduct in the past decade has highlighted that 2.9% of researchers have committed at least one form of scientific misconduct such as text integrity (e.g., plagiarism) and data integrity (e.g., falsification or fabrication, Xie et al., 2021). As per the latest update on the Retraction Watch database, at least 454 articles that aim to understand various aspects of COVID-19 research were retracted due to different forms of scientific misconduct (Retraction Watch 2024). Fanelli (2009) reported that 2% of scientists admitted their involvement in data fabrication or falsification in their careers. A study which compared 121 assays which help to understand toxicological effects on laboratory animals, found only 57% of them could be reproduced (Gottmann et al., 2001). Thus, journals, funding agencies as well as researchers have addressed the reproducibility issues with comprehensive standards for experiment design, data analysis and reporting results (Landis et al., 2012; Rooney et al., 2016; Schmied et al., 2024). The guiding principles of research integrity not only incorporate transparency, and reproducibility for scientific issues, but also accountability for study subject protection, and interpretation of study results objectively (Robishaw et al., 2020).

Apart from typical scientific misconduct (plagiarism, falsification or fabrication) which is considered a serious deviation from research conduct, there are other subtle practices (e.g., conflict of interest, suppression of negative results) that also breach research integrity (Shaw & Erren, 2015). There are various rules and regulations formulated around the globe at the national, international and institutional levels to ensure research integrity; however, researchers often encounter issues which require them to make decisions as per ethical guidelines rather than blindly following the rules (Geller et al., 2010; Shaw & Erren, 2015). This, in the long run, helps them to prevent any deviations from standard research ethics and integrity practices rather than simply reacting to them (Davis et al., 2007).

Research Ethics and Integrity Office (RIO) as a Measure to Tackle Scientific Misconduct

In 1981, the Investigations and Oversight Subcommittee of the House Science and Technology Committee in the United States held its first hearing on scientific misconduct as it became an emerging public issue, particularly at research centres. Around twelve cases of scientific misconduct in the USA between 1974 and 1981 came to light which received congressional attention as allegations that the National Institute of Health (NIH), universities, and institutes did not adequately address the scientific misconduct cases (Scheetz, 1999). This prompted the U.S. Congress to pass the Health Research Extension Act in 1985 which required all institutions across the United States seeking federal research grants to establish appropriate administrative processes to review any report of scientific misconduct and inform the Secretary about the outcome of the alleged research misconduct where the fraud appears substantial. To implement the law, the Public Health Service (PHS) created the Office of Scientific Integrity (OSI) with the Office of the Director at NIH as well as an Office of Scientific Integrity Review (OSIR) as part of the Office of the Assistant Secretary for Health in 1989 and later consolidated these offices as ORI (Office of Research Integrity) and the Office of Health and Human Services (HHS) in 1992 to address all research misconduct reports and applications (Scheetz, 1999). The functional sphere of ORI includes policy and regulation development, conducting and reviewing investigations, as well as presenting the investigation findings to HHS Departmental Appeals Board (DAB) and conducting training, and capacity-building programmes in academic integrity. To tackle increased incidents of scientific misconduct more comprehensively, China has created an Office of Scientific Research Integrity, but most low and middle-income countries still lack dedicated research integrity offices (Ana et al., 2013).

Function and Responsibility of Research Ethics and Integrity Officer

The function and responsibility of any Research Ethics and Integrity Officer are often considered complex and unique due to the absence of similar analogous positions in the higher education ecosystem (Wright & Schneider, 2010). The role of an RIO officer is often considered "quasi-judicial," which primarily includes the responsibility of a regulatory manager, educator, mediator and prosecutor if needed. In the process of addressing a research misconduct allegation, the Research Ethics and Integrity Officer is often responsible for processing the allegation by serving notice to respondents involved in a case about rules and responsibilities, obligations as per government and institutional policies, organising or participating in investigating committee meetings or interviews, conducting an initial analysis of a questioned document or dataset, drafting or reviewing investigating committee reports and maintaining communication with various government agencies. The officer also maintains communication with other relevant offices (e.g., regulatory compliance, finance) if there is a complex case which involves multiple areas such as violation of animal or human ethics regulations protecting whistle-blowers from reprisal acts. Additionally, the office also plays an important role in identifying, and managing conflicts of interest (e.g., financial, intellectual), and research data management for various stakeholders at their respective host institutes to improve public trust in research (Robishaw et al., 2020).

Research Ethics and Integrity: An Indian Perspective

India is committed to upholding logical and rational thinking as a fundamental duty of every citizen, as highlighted in Article 51A (h) of the Indian constitution, and aspires to contribute significantly to global scientific advancement. It also sees science as a way to meet its various sustainable development goals, including providing quality education to all. Many government agencies such as the University Grant Commission (UGC), and the Indian Council for Medical Research (ICMR) have formulated detailed policy documents to address challenges in academic integrity (Bhargava et al., 2019; Patwardhan et al., 2020). However, the oversight mechanisms to ensure academic integrity and compliance vary significantly across different organisations, leading to implementation challenges. Moreover, the specific challenges faced by Indian academics are not well documented. A multicentric and cross-sectional study focused on postgraduate students and faculties in selected medical colleges in India highlighted that participants needed more knowledge to avoid plagiarism (Raj et al., 2021). Cobey et al. (2019) indicated that many Indian researchers publish in predatory journals as they mistakenly identify them as authentic peer-reviewed journals. Thus, a lack of understanding about academic integrity and the absence of a dedicated research ethics and Integrity office at the institutional level not only hampers effective implementation and interventions, but may also result in increased incidents of research misconduct (intentional and unintentional) or lapses in regulatory compliance. Thus, the University Grants Commission (UGC) recommended the establishment of research integrity offices at Indian higher education institutes to prevent incidents of academic malpractice (Jayaraman, 2020).

The Bangalore Life Science Cluster (BLiSC) comprises multiple institutes, each with different reporting structures. Two of them, the Institute for Stem Cell Science and Regenerative Medicine (iBRIC-inStem) and the National Centre for Biological Sciences (NCBS-TIFR) in Bangalore, India are known for cutting-edge, internationally recognised research in biological sciences. They share a common campus, and common research facilities, with cross-campus collaborations between faculties, scientists and students. Institutes within BLiSC also have research ethics and integrity policies, awareness policies about research misconduct and for protecting intellectual properties. To meet both reporting requirements, facilitate training, and meet the requirements from UGC to provide a focused course on research and publication ethics, BLiSC established India's first dedicated research ethics and integrity office (RIO). In this chapter, we aim to examine the process of creating a RIO and address relevant challenges and discuss respective solutions. Furthermore, we hope the chapter will also help to raise awareness about the requirement and steps for building a research ethics and integrity.

Formation and Function of Multi-Institute Research Ethics and Integrity Office

Process of Building Research Ethics and Integrity Office

The primary strategy for establishing a multi-institute Research Ethics Integrity office was based on identifying the need, setting up a timeline for activities and creating a process for capacity-building programmes to improve the reproducibility and transparency of the research conducted on campus. The institute leadership envisioned the office to also serve as a liaison between the host institute and various government agencies regarding research ethics and integrity issues. The following are the essential steps which were pivotal in setting up the office in early 2022.

Step I: To build the multi-institute office, a top-down approach was followed. The institute leadership (director, deans, administrative and finance heads) frequently met to discuss the feasibility of a multi-institute research ethics and integrity office, relevant administrative policies, financial commitments and modalities for staff recruitment for the office. This was particularly important as two primary participating institutes are affiliated with separate ministries of Govt. of India with a unique set of rules and regulations. Thus, initial discussions played a key role in designing the overarching role and function of the RIO.

Step II: Research ethics officers were recruited with the help of a selection committee consisting of members from both participating institutes. Upon joining, ethics officers regularly met with institute leadership to identify short-term and long-term goals/priorities as well as set a realistic timeline to achieve these goals. Additionally, ethics officers also met various stakeholders to understand their expectations from the office.

Step III: A four-member (e.g., Head of Research, Dean of Academics) multi-institute faculty advisory committee was formed which closely advises and oversees the functions of the RIO. The research ethics officer and committee members meet weekly during the initial formation phase of the office and later at least once a month to discuss all relevant matters from each participating institute. The meeting minutes get shared with all committee members and RIO personnel. This structured form of communication between the advisory committee and RIO personnel becomes crucial to identify tasks that need to be completed, and the extent of any modifications, if required. Being a multi-institute RIO, all processes (e.g., functional, administrative) were designed as per the extant rules and regulations of each participating institute.

Step IV: RIO personnel frequently met with various stakeholders (faculties, students, researchers) in person at regular intervals to understand their views, comments, and suggestions about various relevant interventions, policies and processes. These meetings helped RIO to fine-tune its processes.

Functional Areas of Research Ethics and Integrity Office

The RIO has defined its overall functional sphere based on the need assessment exercise at the initial phase of office establishment and based on suggestions from various stakeholders. The RIO at BLiSC is primarily working with three areas: training and capacity building, policy interventions and documentation.

A. Training and capacity building: RIO is actively involved in capacity building and training programmes for all stakeholders (e.g., doctoral students, project personnel, research associates, trainees, and visitors) on campus. The BLiSC campus receives visitors (researchers, scientists, interns, and project fellows) from all over India and abroad who are in different career stages and might have diverse levels of understanding about basic academic integrity issues as well as relevant institute policies. Thus, the research ethics and integrity officer conducts a one-hour-long monthly ethics seminar for visitors to familiarise them with basic areas of academic integrity, and facilities available at each institute as well as to highlight institutional policies to address research misconduct. After the seminar, each visitor participates in a multiple-choice question-based exercise which helps RIO to assess the effectiveness of the seminar. The RIO is also actively involved in conducting "Research and Publication Ethics", a 2-credit course, mandatory for all doctoral students to fulfil their doctoral degree requirement. The research ethics and integrity officer delivered this course along with other colleagues from campus covering various relevant topics such as the prevention of plagiarism, conflict of interest, the peer review process, and research data management using traditional lectures, group discussions, guest lectures as well as online games (e.g., Age of Integrity) developed by University of Wollongong, Dubai and European Network for Academic Integrity (2024). Since its inception in 2022, more than 2000 visitors and 220 doctoral students have been trained through monthly ethics seminars and research and publication ethics courses. RIO also received positive responses from attendees through an online anonymous course feedback survey. Here are two student comments from the survey which highlight course quality: "The way science is done is very important. This course teaches you to distinguish the good and poor ways of doing science."; "The course helped me understand the importance of research ethics. Most importantly, the discussions brought out various insights of ethics from everyone's perspective."

B. Policy formulations and intervention: In consultation with institute leadership and the faculty advisory committee, the RIO formulated policies and protocols for using tools for academic integrity (e.g., text and image integrity) to help uphold the academic integrity of research activities at the institutes. The Research Ethics and Integrity Officer acts as the lead person to procure text and image integrity tools and conduct user training workshops, creating a standard operating protocol for the usage of tools and helping users in troubleshooting as a liaison between users and vendors. The Research Ethics and Integrity Officer also makes available all relevant institute, national and international policies and guidelines on the RIO website as a resource for various stakeholders.

C. Documentation: RIO also created process and space for publication data archiving to improve the transparency and reproducibility of research conducted at the institutes. The primary aim of the documentation process is quick retrieval of the publication data when required. The RIO created a standard operating protocol, policy and publication checklist to facilitate publication data archive. The corresponding author fills up the checklist with information about the article name, journal name, text and image integrity checks, and various regulatory compliance permission numbers. The research ethics and integrity officer creates an online folder for data deposition in the institute-specific online server and shares the link with the corresponding author. The corresponding author and other co-authors arrange the data per each figure, table, supplementary materials, and metadata. The research ethics and integrity officer checks the data structure and finally locks the data deposition folder (see Fig. 2.1 below).

In addition to the above-mentioned functions, the RIO has been actively involved in various outreach and knowledge-sharing activities outside of BLiSC campus. For example, the RIO actively participated in delivering lectures (online and in person) and conducted research ethics workshops for other Indian academic institutes (e.g., TIFR Mumbai, Wildlife Institute of India, Dehradun), helped academic institutes to identify appropriate academic integrity tools (e.g., ACTREC, Mumbai) and conducted collaborative workshops (e.g., ethics of grant writing at Indian Institute of Science, Bangalore). The RIO has also delivered online lectures to create awareness among undergraduate students across India by participating in online undergraduate

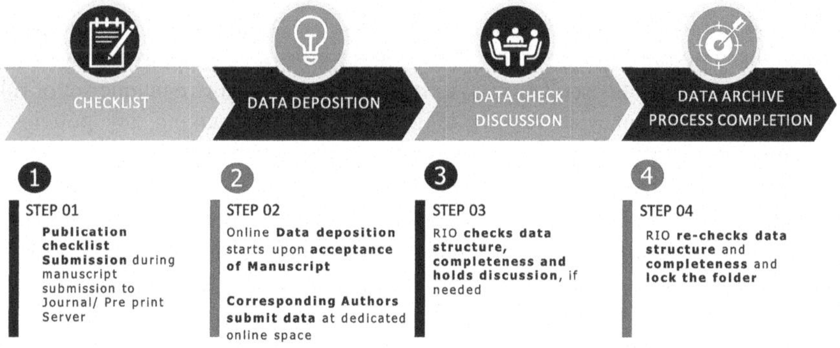

Fig. 2.1 Stages of publication data archiving process to improve the reproducibility and transparency in scientific research

lecture series organised by BLISC, through participation in the flagship lecture online series known as "Science Setu".

The Research Ethics and Integrity Office also engaged in research on various aspects of academic integrity. In 2022, the office received a competitive grant from the DBT/Wellcome Trust India Alliance to understand the current status of academic integrity and regulatory compliance at Indian Institutes. As a part of the project, RIO collaborated with the regulatory compliance office on campus and launched a nationwide online survey to understand gaps and challenges in academic integrity and regulatory compliance at life science departments at Indian institutes, created a database for teaching materials for educators which will help them in teaching various relevant topics of academic integrity and regulatory compliance. Furthermore, an in-person two-day training workshop was also organised at iBRIC-inStem in February 2024 in which 21 life science faculty working at higher education institutes in 13 different Indian states also participated. The anonymous workshop feedback from participants highlighted the need for research ethics and integrity workshops for academicians. Here are two examples of feedback received from workshop participants, feedback 1: "Through this workshop, I could learn an ethical, mindful, and balanced approach to biological research. It has also helped me develop ethics for using model organisms. This workshop also enabled me to help my institution build policies on academic integrity and regulatory compliance."; feedback 2: "The workshop made me realise how important academic regulations and ethical issues are in research methodology. Universities do not have such regulatory setups which I believe need to be created now for transparency in research communication. The overview of AI-based tools and software for image integrity was a new experience for me. The online games on research ethics are certainly a good takeaway from the workshop that I will use in my academic teaching.".

Unique Features of Multi-Institute RIO

The multi-institute research ethics and integrity office facilitates the exchange of ideas and intellectual inputs from large and diverse groups of stakeholders which makes the functions, and policies of RIO more holistic and impactful. A distinct feature of multi-institute collaborative effort is acknowledging a unique set of rules and regulations from each participating institute while making any policies or designing any intervention. Thus, RIO regularly meets not only the faculty advisor committee members, but also the administrative team at each institute to ensure all policies comply with the participating institute norms. Additionally, multi-institute participation also allowed RIO to utilise the combined financial resources to procure advanced and cutting-edge research integrity tools for text and image integrity checks. The use of various tools enables scientists and researchers working within BLiSC campus to avoid any unintentional errors during the reporting and publication of their results. The financial commitment towards procurement of integrity tools was estimated as per the number of users from each participating institute.

There were other facilities such as a library, research facilities which are common for participating institutes, thus functioning costs for these facilities were divided equally.

Challenges to Build Research Ethics and Integrity Office

The number of research integrity officers is usually limited per institute with inadequate formal training opportunities for them. Often research integrity offices experience a high turnover of personnel. In the USA, a research integrity officer stays on the job for an average of about 3 to 5 years (Wright & Schneider, 2010). This hampers effective knowledge transfer from predecessor officers to newly joined personnel as well as limited scope to gain experience about the complex job role of a research integrity office. Many misconduct cases handled by RIO are litigious, thus mishandling any cases results become serious problems for the office as well as the Institutions. These situations adversely impact the overall function of the RIO. To resolve this challenge, institutes can envision the role of an RIO into the larger scope of the Institute mandate and empower the research ethics and integrity officer with a competitive salary, adequate supervision as well as career-building opportunities.

In spite of various existing rules, regulations and guidelines for research ethics and integrity, a lack of detailed standard operating protocols, and policies hamper appropriate implementations of research integrity regulations (Shaw & Erren, 2015). The various stakeholders, particularly junior researchers at each academic institute come from diverse backgrounds without clear knowledge about the procedure to raise scientific misconduct-related concerns. Furthermore, whistleblowers are often concerned about retaliation from the offender. Thus, RIO at academic institutions needs to have clearly defined procedures for reporting research misconduct, various steps and timelines of the investigations, as well as whistleblower protection policies publicly available on their office website. Furthermore, appropriate enforcement of research integrity policies, as well as adequate power to the research integrity office to initiate misconduct investigations are also essential to enhance the impact and functionality of any RIO (Shaw & Erren, 2015).

The limitations of resources may also impact the formation of a dedicated research ethics and integrity office. Dickens and Cook (2003) highlighted that the lack of resources significantly constrains the functionality and impact of bioethics committees. The resource limitations also pose significant challenges in building a dedicated research ethics and integrity office, organising regular ethics committee meetings/quorums, procuring software for text and image integrity checks, and hiring of support staff for the office. A multi-institute research ethics and integrity office can play a pivotal role in addressing this challenge. This approach allows institutes to utilise combined resources in line with user numbers from each institute. The multi-institute approach helps to reduce the financial burden on individual institutes where combined resources allow researchers to explore the state of art tools in research ethics and integrity and make them available to various stakeholders on campus. Additionally, funders can also make resources (e.g., competitive funds, training opportunities) for research ethics and integrity officers for their capacity building.

Research misconduct is a global issue and many countries in the global north have successfully been able to build various networks of ethics professionals. These networks help new RIO and ethics professionals in capacity building and knowledge sharing. Similar networks are not very common among countries in the global south, which hampers newly recruited research ethics and integrity officers, particularly in the global south, to understand the nuances of academic integrity, as well as various key administrative procedures associated with RIO function. Therefore, new ethics professionals can communicate with experts and members in different existing professional networks such as the International Centre for Academic Integrity (ICAI, 2024) which has various consortia in USA and Canada, European Network for Academic Integrity (ENAI, 2024) and European Network of Research Integrity Offices (ENRIO, 2024) which provides networking platforms for ethics professionals from Europe and beyond. There are also country-specific networks such as the UK Research Integrity Office (UKRIO, 2024), the Netherlands Research Integrity Network (NRIN, 2024), Australasian Academic Integrity Network (AAIN, 2023) for ethics professionals working in Australia and New Zealand. Senior professionals from these networks can significantly help institutes build their RIO as well as relevant procedures including capacity-building programmes. The RIO officer at BLiSC searched many of these professional networks as well as made direct contact via email with ethics officers working at various academic institutes in Austria, Australia, Germany, UK, USA. The ethics officer from each of these overseas institutes/universities informed the BLiSC RIO officer about how their office functions, tools they commonly use, relevant processes, challenges and their probable solutions over online meetings. Information obtained through those meetings was very helpful while making RIO's own processes and guidelines.

Conclusion

Research misconduct remains a major concern among scientists, funders, and policymakers, as it not only erodes public trust, but also causes a significant amount of financial loss. More than 10,000 articles were retracted from journals in the year 2023 due to various ethical and regulatory compliance issues (Van Noorden, 2023). This rate of article retraction is unprecedented. Thus, the establishment of research ethics and integrity offices at academic institutes can help to prevent incidents of research misconduct. In this chapter, we have described the process to build a multi-institute research ethics and integrity office (RIO), as well as its function which can significantly contribute to make research conducted at academic institutes more ethical, compliant, transparent and reproducible. We recommend following the process of setting up multi-institute RIO described here as a best practice example of cooperation, and optimised utilisation of limited resources in resource-limited settings.

Acknowledgements We are thankful to the Director, iBRIC-inStem, and Director, NCBS-TIFR for providing support and guidance to establish the Research Ethics and Integrity Office. We would also like to thank DBT/Wellcome Trust India Alliance for the IRMI RM Travel Grant (IA/TG/22/1/7100018) to the first author to facilitate his participation in the ninth European Conference on Ethics and Integrity in Academia (ECEIA) 2023 in Derby, UK which helped the authors in knowledge sharing with other research ethics and integrity officers in UK, Europe and beyond.

References

Ana, J., Koehlmoos, T., Smith, R., & Yan, L. L. (2013). Research misconduct in low-and middle-income countries. *PLoS Medicine, 10*(3), e1001315. https://doi.org/10.1371/journal.pmed.1001315

Australasian Academic Integrity Network (AAIN). (2023). *Welcome*. https://academicintegrity.edu.au

Bhargava, B., Gangakhedkar, R.,Jain, N.C., Mathur, R. (2019). ICMR policy on research integrity and publication ethics. *Retrieved March 10*, 2024, from https://www.icmr.gov.in/icmrobject/uploads/Guidelines/1724844182_icmr_pripe2019.pdf

Cobey, K. D., Grudniewicz, A., Lalu, M. M., Rice, D. B., Raffoul, H., & Moher, D. (2019). Knowledge and motivations of researchers publishing in presumed predatory journals: A survey. *BMJ Open, 9*(3), e026516. https://doi.org/10.1136/bmjopen-2018-026516

Davis, M. S., Riske-Morris, M., & Diaz, S. R. (2007). Causal factors implicated in research misconduct: Evidence from ORI case files. *Science and Engineering Ethics, 13*, 395–414. https://doi.org/10.1007/s11948-007-9045-2

Dickens, B. M., & Cook, R. J. (2003). Challenges of ethical research in resource-poor settings. *International Journal of Gynaecology & Obstetrics, 80*(1), 79–86. https://doi.org/10.1016/S0020-7292(02)00349-1

European Network for Academic Integrity (ENAI). (2024). *Homepage*. https://www.academicintegrity.eu/wp/

European Network of Research Integrity Offices (ENRIO). (2024). *Homepage*. https://www.enrio.eu

Fanelli, D. (2009). How many scientists fabricate and falsify research? A systematic review and meta-analysis of survey data. *PloS One, 4*(5), e5738. https://doi.org/10.1371/journal.pone.0005738

Geller, G., Boyce, A., Ford, D. E., & Sugarman, J. (2010). Beyond "compliance": The role of institutional culture in promoting research integrity. *Academic Medicine, 85*(8), 1296–1302. https://doi.org/10.1097/ACM.0b013e3181e5f0e5

Gottmann, E., Kramer, S., Pfahringer, B., & Helma, C. (2001). Data quality in predictive toxicology: Reproducibility of rodent carcinogenicity experiments. *Environmental Health Perspectives, 109*(5), 509–514. https://doi.org/10.1289/ehp.01109509

Gupta, A. (2013). Fraud and misconduct in clinical research: A concern. *Perspectives in Clinical Research, 4*(2), 144–147. https://doi.org/10.4103/2229-3485.111800

International Centre for Academic Integrity [ICAI]. (2024). *Homepage*. https://academicintegrity.org

Jayaraman, K. S. (2020, October 11). *Research integrity offices to check academic malpractice: UGC*. Nature India. https://doi.org/10.1038/nindia.2020.159

Landis, S. C., Amara, S. G., Asadullah, K., Austin, C. P., Blumenstein, R., Bradley, E. W., Crystal, R. G., Darnell, R. B., Ferrante, R. J., Fillit, H., Finkelstein, R., Fisher, M., Gendelman, H. E.,

Golub, R. M., Goudreau, J. L., Gross, R. A., Gubitz, A. K., Hesterlee, S. E., Howells, D. W., Huguenard, J., Kelner, K., Koroshetz, W., Krainc, D., Lazic, S. E., Levine, M. S., Macleod, M. R., McCall, J. M., Moxley, R. T., III, Narasimhan, K., Noble, L. J., Perrin, S., Porter, J. D., Steward, O., Unger, E., Utz, U., & Silberberg, S. D. (2012). A call for transparent reporting to optimize the predictive value of preclinical research. *Nature, 490*(7419), 187–191. https://doi.org/10.1038/nature11556

Netherlands Research Integrity Network (NRIN). (2024). *Homepage*. https://www.nrin.nl

Patwardhan, B., Desai, A., Chourasia, A., Nag, S., & Bhatnagar, R. (2020). *Guidance document: Good academic research practices*. University Grants Commission. https://www.ugc.gov.in/e--book/UGC_GARP_2020_Good%20Academic%20Research%20Practices.pdf

Raj, J. P., Venkatachalam, S., Amaravati, R. S., Siby, N., Oommen, A. M., Jose, J. E., George, M., Ramraj, B., Penumutsa, V. S., Bodda, D. R., Rajad, R., Reshmi, R., Inbaraj, L. R., Rajiv, S., Bhandare, B., Aiyappan, R., Mathew, P., Saleem, A. K., Shetty, R., Bhuvaneswari, G., Kumar, S. T., Prasanna, B. L., & Manderwad, P. G. (2021). Extent of awareness and attitudes on plagiarism among post-graduate resident doctors and junior medical faculty in India: A cross-section. *BMJ Open, 11*(6), e046904. https://doi.org/10.1136/bmjopen-2020-046904

Retraction Watch. (2024). Retrieved November, 2024, fromhttps://retractionwatch.com/

Robishaw, J. D., DeMets, D. L., Wood, S. K., Boiselle, P. M., & Hennekens, C. H. (2020). Establishing and maintaining research integrity at academic institutions: Challenges and opportunities. *The American Journal of Medicine, 133*(3), e87–e90. https://doi.org/10.1016/j.amjmed.2019.08.036

Rooney, A. A., Cooper, G. S., Jahnke, G. D., Lam, J., Morgan, R. L., Boyles, A. L., et al. (2016). How credible are the study results? Evaluating and applying internal validity tools to literature-based assessments of environmental health hazards. *Environment International, 92*, 617–629. https://doi.org/10.1016/j.envint.2016.01.005

Scheetz, M. D. (1999). Office of Research Integrity: A reflection of disputes and misunderstandings. *Croatian Medical Journal, 40*(3), 321–325.

Schmied, C., Nelson, M. S., Avilov, S., Bakker, G. J., Bertocchi, C., Bischof, J., et al. (2024). Community-developed checklists for publishing images and image analyses. *Nature Methods, 21*(2), 170–181. https://doi.org/10.1038/s41592-023-01987-9

Shaw, D. M., & Erren, T. C. (2015). Ten simple rules for protecting research integrity. *PLoS Computational Biology, 11*(10), e1004388. https://doi.org/10.1371/journal.pcbi.1004388

Smith, R. (2006). Research misconduct: The poisoning of the well. *Journal of the Royal Society of Medicine, 99*(5), 232–237. https://doi.org/10.1177/014107680609900514

UK Research Integrity Office (UKRIO). (2024). *Homepage*. https://ukrio.org

Van Noorden, R. (2023). More than 10,000 research papers were retracted in 2023—a new record. *Nature, 624*(7992), 479–481. https://doi.org/10.1038/d41586-023-03974-8

Wright, D. E., & Schneider, P. P. (2010). Training the research integrity officers (RIO): The federally funded "RIO boot camps" backward design to train for the future. *Journal of Research Administration, 41*(3), 99–117.

Xie, Y., Wang, K., & Kong, Y. (2021). Prevalence of research misconduct and questionable research practices: A systematic review and meta-analysis. *Science and Engineering Ethics, 27*(4), 41. https://doi.org/10.1007/s11948-021-00314-9

Chapter 3
Presenting a Student Academic Integrity Champion Model: A Proactive Step towards Campus-Wide Efforts in Building a Culture of Integrity

Zeenath Reza Khan and Neha Hemnani

Abstract Academic integrity, crucial in education, research, and scholarship, faces challenges globally due to student misconduct, including cheating and plagiarism. Studies indicate widespread instances of such behaviour, with over 60% of students admitting to engaging in academic misconduct. Various factors influence students' attitudes toward integrity, affecting their behaviour during assessments. Reactive measures like using technology for detection and penalties often exacerbate the gap between institutions and students. Conversely, proactive strategies such as training modules, policy reviews, and designing assessments differently have been advocated to foster a culture of integrity on campuses.

Students play a pivotal role in promoting academic integrity; yet engaging them amidst contemporary challenges remains difficult. Efforts to involve students in integrity-related initiatives often encounter barriers. Addressing these challenges necessitates a comprehensive approach involving all stakeholders. This paper presents the development of a Student AI Champion Model (SAIC Model) based on a centre of excellence and a campus's proactive efforts in the Middle East. The model draws on student-centred learning principles and engagement frameworks, emphasising collaboration among stakeholders. Key elements include policy reviews, grassroots implementation, and research activities involving students.

The SAIC Model offers a holistic approach to cultivating integrity, encompassing awareness campaigns, policy revisions, student ambassador programs, and support mechanisms for students facing challenges. By documenting the methodology and successful implementation of the model, this chapter aims to share the SAIC model with other campuses seeking to strengthen their culture of academic integrity.

Keywords Students champions · Partnership · Policies · Model · UAE · Middle East

---

Z. R. Khan (✉) · N. Hemnani
University of Wollongong in Dubai, Dubai, UAE
e-mail: zeenathkhan@uowdubai.ac.ae; nhh996@uowmail.edu.au

M. Davis, C. Baptista (eds.), *Ethics and Integrity in Education (Practice)*, Ethics and Integrity in Educational Contexts 10,
https://doi.org/10.1007/978-3-031-95686-7_3

Introduction

Academic integrity is foundational to the ethos of education, research, and scholarship, serving as a crucial pillar that supports the quality of academic work. It encompasses a range of ethical principles, including honesty, fairness, trust, courage, respect, and responsibility (ICAI, 2021). These values ensure that the work produced by students and scholars is original, verifiable, and produced without deception. The importance of academic integrity transcends the boundaries of individual assignments and tests, impacting the credibility of degrees, the reliability of research findings, and the reputation of educational institutions (Hussinger & Pellens, 2017; Parnther, 2020). Moreover, fostering a culture of integrity prepares students not just for academic success but for honourable conduct in their professional and personal lives, promoting a society where truth and ethical behaviour are upheld (Guerrero-Dib et al., 2000).

Globally, academic institutions face significant challenges due to student misconduct, with cheating and plagiarism at the forefront. Recent studies, such as those by the International Center for Academic Integrity (ICAI), indicate that over 60% of students admit to some form of academic dishonesty during their educational careers (ICAI, 2023). This widespread issue not only undermines the value of educational achievements, but also poses a threat to the foundational trust on which scholarly work is built. Factors contributing to this misconduct include the accessibility of digital resources that simplify the act of plagiarism, pressures to excel academically, and often ambiguous policies regarding integrity (Khan & Balasubrmanian, 2022). These challenges necessitate robust solutions that address both the acts of dishonesty and the cultural factors that promote such behaviours, emphasising the need for an integrated approach to promote and uphold the core values of academic integrity across all levels of education.

The remainder of this chapter is organised as follows: in the next section we describe factors that influence academic integrity, followed by current measures and limitations to combat academic dishonesty, the next section looks at proactive efforts by institutions, and the role of students in upholding academic integrity. Section five provides details of how the SAIC model was developed and implemented and how it worked, followed by conclusion.

Factors Influencing Academic Integrity

Various factors significantly influence student attitudes toward academic integrity, as highlighted by researchers such as Vučković et al. (2020). Cultural influences play a pivotal role, as societal norms and values can greatly affect how students perceive and engage with ethical conduct. In some cultures, collaborative work and sharing answers may be seen as a form of mutual support rather than dishonesty, leading to differing views on what constitutes cheating (Farahat, 2022; Zhao et al.,

2022). Additionally, the immense pressure to excel academically can drive students to compromise their ethical standards. This pressure is often exacerbated by competitive education systems and high expectations from family and society (Khan, 2014). Lack of awareness about the policies of academic integrity and the consequences of violating assessment and coursework integrity also contributes (Anohina-Naumeca et al., 2020), as students may not fully understand what actions are considered dishonest or why they are problematic.

These influencing factors manifest in student behaviour during assessments in various ways. Cultural norms and the misunderstanding of academic integrity guidelines can lead to an increase in collaborative cheating or unauthorised sharing of answers during exams. The intense pressure to succeed may push students to engage in plagiarism or other forms of cheating to secure higher grades or meet the high expectations set upon them. Moreover, the lack of comprehensive education on the importance of academic integrity (Khan et al., 2022b) means students might not recognize the severity of their actions or the potential long-term consequences, such as damage to their academic record and future career opportunities. This combination of factors necessitates educational interventions that not only clarify what constitutes ethical behaviour, but also build an environment where integrity is valued and rewarded.

Current Measures and Their Limitations

Institutions traditionally employ a variety of reactive measures to combat academic dishonesty, focusing primarily on detection and punishment (Awad et al., 2013). Text similarity detection tools, such as software that scans student work for similarity, are widely used to identify instances of copying and paraphrasing without proper attribution (Khan, 2013). These tools are integral to the administrative process of enforcing academic integrity policies. Alongside these technical solutions, punitive measures are also prevalent. Depending on the severity of the misconduct, penalties can range from failing grades on specific assignments to suspension or expulsion from the institution. Such measures are intended to serve as deterrents, underlining the seriousness with which institutions treat breaches of academic integrity.

However, the reliance on such reactive measures has been critiqued for potentially exacerbating the relational gap between students and educational institutions. One of the primary criticisms is that these approaches often fail to address the underlying issues that lead to academic dishonesty (Khan et al., 2020a, 2022b). Instead of fostering a learning environment where ethical decision-making is encouraged, an overemphasis on detection and punishment can create an adversarial atmosphere. Students may come to view these measures as mere obstacles to be circumvented, rather than as part of a broader educational ethic (Norris, 2019). This can lead to a lack of trust and diminished student engagement, as students feel they

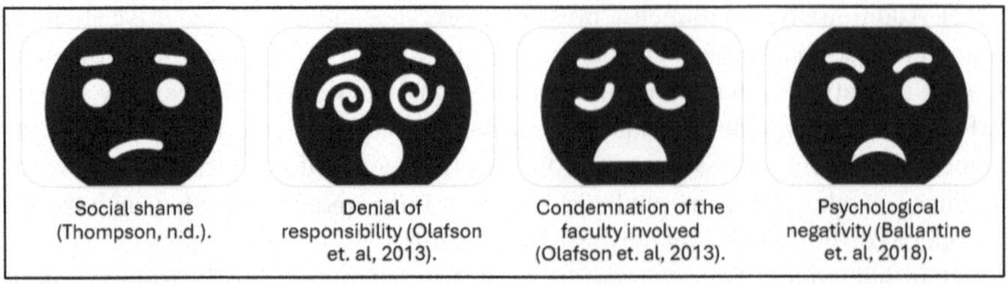

Fig. 3.1 Impact of confrontations, investigations and inquiry on students (adapted from Khan et al., 2020a, 2020b)

are constantly under suspicion and that the educational relationship is punitive rather than supportive.

Moreover, the use of punitive measures can sometimes result in negative outcomes that extend beyond the immediate academic consequences. For instance, according to a study by Khan et al., 2020b, students who are caught and punished for plagiarism may experience significant academic and psychological distress, which can affect their overall educational trajectory and self-esteem (see Fig. 3.1 adapted from Khan et al., 2020b). Such outcomes not only hinder the educational process, but can also leave lasting impacts on student lives. Additionally, heavy reliance on either text matching and artificial intelligence content generation detection tools can lead to false accusations where common phrases or technical terms trigger detection algorithms, further straining the student-institution relationship (Freedman, 2004; Weber-Wulff et al., 2023; Zwagerman, 2008). This critique emphasises the need for a more balanced approach that includes preventive and educational strategies alongside reactive measures, aiming to cultivate an intrinsic respect for academic integrity rather than a fear of punitive consequences.

Proactive Strategies for Promoting Integrity

Proactive strategies for promoting academic integrity have gained traction for their effectiveness in nurturing a culture of ethical behaviour in educational settings. Notably, comprehensive training modules designed to educate students about the principles of academic integrity and the specifics of their institution's policies have proven particularly beneficial (Benson et al., 2019). These modules often include interactive components such as scenarios and quizzes that help students understand the practical application of integrity in their academic work.

Additionally, regular reviews and updates of academic integrity policies ensure that guidelines keep pace with changing educational environments and technologies, thereby reducing ambiguities that can lead to misconduct (Akbar & Picard, 2019). Furthermore, innovative assessment designs that emphasise unique, personalised assignments or open-book exams can reduce opportunities for dishonest behaviours by making cheating less feasible or less rewarding.

Researchers like Khan, Priya, and Tuffnell (2022a) and Stoesz et al. (2019) have highlighted the success of such approaches in various studies, noting substantial improvements in student attitudes towards integrity following the implementation of these strategies.

The Role of Students in Upholding Academic Integrity

Students are central to the successful implementation and sustainability of academic integrity policies within educational institutions that can foster a culture of integrity. Their active participation in promoting honesty and fairness is crucial, as it helps create a self-regulating community where ethical practices are the norm rather than the exception (John et al., 2021). When students understand and value the principles of academic integrity, they are more likely to act as ambassadors of these values, influencing their peers and contributing to a positive educational environment. Moreover, involving students in the development and enforcement of integrity policies can lead to more effective and relevant approaches, as students are likely to offer unique insights into the academic and social pressures that could lead to misconduct (Çelik et al., 2022).

However, engaging students in integrity initiatives often presents significant challenges. According to Khan et al. (2019), one of the primary barriers is the lack of awareness and understanding about what constitutes academic dishonesty and its implications. Many students do not fully comprehend the breadth of behaviours that constitute academic misconduct, nor are they aware of the consequences these actions can have on their academic and professional lives (Diego, 2017). Furthermore, there can be a perceptible disconnect between faculty and students on the importance of academic integrity, with students sometimes viewing these rules as arbitrary or only applicable to certain contexts.

Additionally, there is often a cultural barrier where norms from various backgrounds may conflict with the principles upheld by educational institutions, leading to misunderstandings and resistance to standard policies. To overcome these barriers, it is essential for institutions to implement comprehensive educational programmes that not only inform students about the rules, but also engage them in discussions about the ethical reasons behind these standards. Creating opportunities for students to take leadership roles in integrity initiatives, such as peer mentoring programs or student-led workshops, can also enhance engagement and foster a deeper commitment to upholding integrity across the campus.

Researchers and academic leaders recognise the critical role that students play as champions in fostering a culture of integrity. However, the literature suggests that solutions must be holistic, involving all stakeholders to be truly effective. McCabe and Pavela (2000) and Hendershott et al. (2000) emphasise the importance of a collaborative approach that includes not only students, but also faculty and administration. Eury and Treviño (2019) further support this view.

To operationalise this inclusive approach, Packalen and Rowbotham (2022) have explored strategies that actively incorporate student perspectives. They advocate for a broad-based dialogue where students contribute their insights on improving the culture of integrity within academic settings. This includes discussions on various dimensions such as student and staff attitudes, pedagogical methods, assessment practices, grading systems, and effective time management. Their work underscores the importance of considering the student voice in developing and implementing policies and practices that enhance the academic integrity landscape across educational institutions.

However, there remains a gap in the literature on how campuses can engage students to become champions of academic integrity. This chapter outlines the development of a student champion model based on a practice case study from a campus in the Middle East and a centre hosted at the campus.

Development of the Student AI Champion Model (SAIC Model)

Premise to the Case Study

The case study was a national-level centre of excellence on academic integrity and its host campus, a western university's offshore campus in the Middle East. A comprehensive longitudinal study by Khan, Priya, and Tuffnell (2022a) detailed the campus's dedicated initiatives to foster a culture of integrity through a holistic approach over a decade. This study outlined the evolution of these efforts, starting with the faculty's move to involve the campus in the International Day of Action Against Contract Cheating in 2016. This involvement was driven by the urgent need to launch awareness campaigns targeting contract cheating, as further elaborated by Khan, Hemnani, et al. (2020a). Khan, Hemnani, et al. (2020a) documented the progressive efforts to engage students as organising members in various events, detailing a systematic approach to involve them incrementally at a western campus in the Middle East. However, as highlighted by Khan et al. (2019), students who stepped forward to participate encountered numerous obstacles. Engaging students to become champions of academic integrity on campus was faced with multiple barriers that could undermine their involvement and effectiveness. To gain a deeper insight into these challenges, informal, open discussion sessions were conducted with student volunteers as part of consented workshops where students actively and explicitly chose to engage. These discussions aimed to explore and better understand the perceived barriers faced by students as illustrated in Fig. 3.2, adapted from Khan and Wallace (2022) and discussed below:

- Peer Pressure: Fear of social isolation or ridicule from peers for upholding standards that others may disregard.
- Perceived Double Standards: Cynicism and distrust arise when students observe inconsistencies in faculty or administrative practices regarding integrity.
- Fear Factor: Concerns about retaliation or misunderstandings about the role of an integrity advocate can deter student participation.
- Missing Personal Connection: Lack of meaningful relationships with faculty or administrators may leave students feeling undervalued and reluctant to engage.
- Value Discrepancy: Some students may not see the relevance or importance of the role of integrity champion to their personal or professional goals.
- Lack of Ownership: Students may feel disconnected from the outcomes of integrity policies, diminishing their motivation to contribute.
- Unawareness of Opportunities: Students often are not informed about the opportunities available to advocate for integrity.
- Unfamiliarity with Rules and Norms: A lack of understanding of campus-specific integrity rules and norms can impede effective student involvement (Fig. 3.2).

These points highlight the need for institutions to foster an inclusive, transparent, and supportive environment that encourages student leadership in integrity initiatives.

Theoretical Framework Supporting Model Development

The development of the Student AI Champion Model (SAICM) provides a comprehensive example of how educational theories and student engagement can be effectively harnessed to cultivate a culture of academic integrity. Inspired by a body of

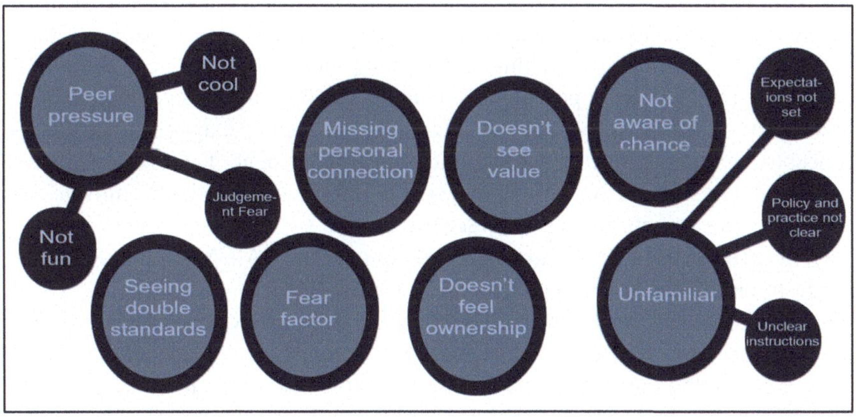

Fig. 3.2 Barriers and challenges that hinder students engaging and volunteering (adapted from Khan & Wallace, 2022)

literature that emphasises active learning and student-centred approaches, the model incorporates several theoretical frameworks that serve as its foundation.

According to Pittaway (2012), engaging students as active learners enhances their involvement and commitment to the educational process. This philosophy aligns with the principles of Student-Centred Learning (SCL), which are rooted in constructivist theory and self-determination theory as discussed by Lee and Hannafin (2016). These theories advocate for an educational environment where students are seen not just as recipients of information but as integral drivers of their own learning journey.

Building on these educational theories, the campus and centre initiated a series of structured activities and engagements to address and overcome various barriers and challenges to academic integrity. The first significant step was at the campus in 2017, a policy review which involved students directly through focus group discussions alongside other stakeholders. This inclusive approach ensured that the policies developed were reflective of the needs and perspectives of the entire academic community, particularly the students.

In 2018, the campus moved towards grassroots implementation of these revised policies. A notable activity was organising debates among students regarding the proposed policy changes. The outcomes of these debates were then presented to the custodian of the policy. This process not only empowered students by giving them a voice in the policy-making process, but also ensured that their insights and concerns were considered at the highest levels of decision-making.

Between 2018 and 2023, the campus and the newly formed national-level centre of excellence intensified their focus on research activities that engaged students from the campus and other campuses as active participants. These activities included internships, assistantships, and opportunities for students to facilitate data collection, co-author presentations, and co-design studies. Such engagements provided practical experiences that reinforced the theoretical aspects of academic integrity and research ethics (John et al., 2021; John et al., 2023).

The development process of the model further drew on the CARE learning design framework, as described by Kostoulas-Makrakis and Makrakis (2020), which emphasises making teaching and learning more student-driven. The CARE learning design framework provides a structured approach to the development of the Student AI Champion Model by focusing on key pedagogical processes: Conceptualization, Activation, Reflection, and Engagement. This framework was instrumental in shaping the various stages of developing the model to cultivate academic integrity among students as described below:

Conceptualization served as the foundation for the SAICM. In this phase, we sought to clarify the key components of academic integrity, including the 'what', 'why', 'who', 'when', and 'where'. This involved defining the scope and aims of the model, identifying the stakeholders (primarily students, faculty, and administrative staff), determining the appropriate timing for implementation, and locating where the model would be most effectively engaged within the campus ecosystem. This stage ensured that the SAICM was thoughtfully aligned with the institution's values and objectives regarding academic honesty.

Activation involved transitioning from the abstract design of the SAICM to concrete learning activities. These activities provided opportunities for students to engage in meaningful dialogue, reflection, and engagement around the topics of integrity and ethics. Students were invited to volunteer and then promoted to lead, brainstorm and organise the initiatives taken to organise awareness campaigns and events across host and other campuses, empowering students and giving them confidence. Workshops, simulations, and discussions activated the student's learning processes and encouraged them to take part as participants or as organisers actively in the promotion and maintenance of academic integrity on campus. It was about moving from theory to practice, where students could actively participate in the shaping and implementation of integrity policies.

Reflection is integral to the CARE framework and the SAICM, as it promotes continuous contemplation of the values, thoughts, feelings, habits, and experiences concerning academic integrity. Students were encouraged to reflect on their personal academic practices, compare them with alternative approaches, and understand the broader implications of academic misconduct. This reflection helped foster a personal commitment to integrity, as students examined how their actions aligned with the community's values and their own.

Engagement in the context of the SAICM refers to the active application of learned principles into practice. It is not just about acquiring knowledge or developing understanding; it is about merging this knowledge and critical consciousness with action. By involving students in the process of policy formation, research activities, and peer-led initiatives, the SAICM went beyond educating about integrity. It provided a platform for students to actively advocate for and embody the principles of academic integrity, thereby reinforcing the model through tangible actions and initiatives on campuses.

By employing the CARE learning design framework in developing the SAICM, we ensured that the process was comprehensive, dynamic, and deeply ingrained in the educational experience of students. It enabled a cyclical process of learning and growth that not only informed students about academic integrity, but also empowered them to be its champions in both theory and practice.

Additionally, the model incorporates the three dimensions of engagement—student-student, student-content, and student-faculty—identified by Ahshan (2021) and Leslie (2020). This holistic approach ensured that all aspects of the learning environment were aligned to support a culture of integrity, making the SAICM not just a policy initiative but a sustainable model that actively involved all stakeholders in the educational process. Through these concerted efforts, the campus established a robust framework that not only addresses immediate concerns around academic integrity, but also fosters a long-term culture that values ethical conduct and active student participation.

Implementation of the SAIC Model

The Student AI Champion Model was methodically brought to life through a series of strategically planned actions and engagements designed to integrate students deeply into the fabric of academic integrity for host campus and the centre:

- **Participation in Integrity Events:** Students were invited to partake in the International Day of Action Against Contract Cheating (IDOA) and engage in debates with faculty members, fostering an environment of open dialogue about integrity.
- **Volunteering Opportunities:** The model offered students roles in volunteering at academic events, allowing them to gain a stake in promoting integrity through practical experience.
- **Workshops and Training:** Specialized AIVAS workshops and training sessions (Khan, 2022) were held for volunteers, focusing on the nuances of academic integrity and writing, to enhance their understanding and skills.
- **Policy Involvement:** Students were actively involved in policy debates and forums nationally and internationally, providing them with a platform to voice their perspectives and contribute to policy discussions.
- **Organizing Committee Roles:** Students took on significant responsibilities as members of organising committees for events related to academic integrity, instilling a sense of ownership and leadership.
- **Recognition Ceremonies:** Ceremonies were held to acknowledge and honour the contributions of student champions and those deeply engaged in research, highlighting the value of their commitment.
- **Research Internships and Training:** Opportunities for research internships and further training were provided, linking academic integrity with professional research practice.
- **Scholarly Contributions:** Students were given avenues to author, co-author, and present research findings, as well as participate in focus groups, conferences, and manuscript writing, embedding them further into the academic community.
- **Celebration of Milestones:** The achievements and progress of student champions were celebrated, reinforcing the importance of their role and encouraging continued advocacy for integrity.
- **Mentorship Development:** Senior students who excelled as integrity champions were promoted to mentor roles, enabling them to guide and pass on the baton of responsibility and knowledge to their juniors. This step ensured the sustainability of the integrity culture by fostering a legacy of mentorship and continuous engagement.

Key Elements of the SAICM

As explained in previous sections above, SAICM is structured around a dynamic and progressive framework designed to cultivate a robust academic integrity culture on campus. Figure 3.3 illustrates this model which is cyclic in nature because it is designed to perpetuate a self-reinforcing loop of academic integrity that contributes to its sustainability. It starts with creating awareness and excitement, which attracts volunteers keen on upholding and spreading the values of academic integrity. These volunteers, through their engagement and learning, evolve into champions and researchers, deepening their understanding and commitment. Their advanced involvement leads to the production of scholarly work, which in turn enhances academic discourse and serves as a tangible testament to the value of integrity.

As these seasoned participants step into mentorship roles, they pass on their knowledge and experience to the next generation of students. This mentoring is crucial because it ensures that the values and skills associated with academic integrity are continuously transmitted. New volunteers are thus always entering the cycle, learning from their predecessors, and eventually replacing them as they move on to become mentors themselves or graduate. The cyclical nature of the model ensures that each cohort of students contributes to a perpetual culture of integrity that becomes ingrained within the institutional fabric. This ongoing cycle not only

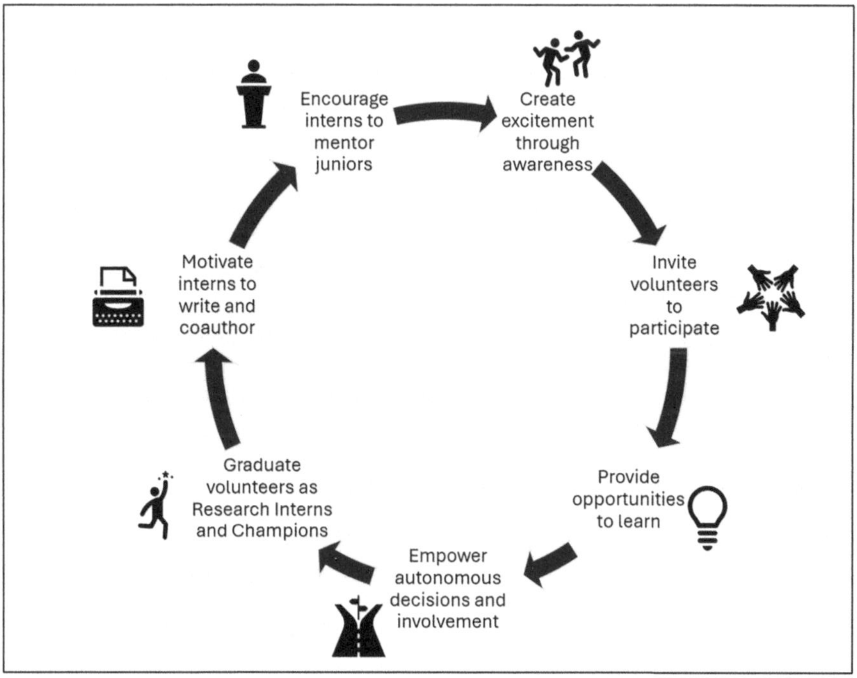

Fig. 3.3 Student AI Champion Model (SAICM)

maintains, but also rejuvenates and adapts the culture of integrity to new challenges and changing academic environments, making the model inherently sustainable.

Impact and Conclusion

Over the span of 7 years, the implementation of the Student AI Champion Model has had a profound impact on the host campus and the centre of excellence for academic integrity. With the establishment of a dedicated student board and an active student committee at the centre that includes 13 student champions engaged throughout the year, the initiative has thrived. This robust student-led governance structure has been instrumental in successfully organising over 20 events and programs, ranging from international conferences to awareness roadshows, deeply embedding integrity within the campus culture. The ripple effect of these initiatives has been remarkable, with a total of 77 ambassadors emerging across various other campuses, each carrying forward the mantle of integrity in their respective domains. These ambassadors have collectively facilitated over 20,000 unique experiential engagements that have touched the lives of not just students, but also faculty, staff, and parents, thereby weaving the principles of academic integrity into the very fabric of the educational community. The breadth and depth of this impact underscores the transformative power of a sustainable, student-driven approach to upholding academic integrity.

The Student AI Champion Model (SAICM) has demonstrated remarkable benefits and successes, creating a robust culture of academic integrity that resonates throughout the host campus and within the ambassadors at the centre. Its success lies in its cyclic nature, which encourages sustainability and continual renewal of commitment to integrity principles. The model's student-centred approach has empowered learners to take active roles in promoting ethical practices, thereby fostering a sense of ownership and pride in upholding these values. The establishment of a centre of excellence with a student board and active committee has been central to these efforts, successfully orchestrating numerous events and engaging a wide audience, creating a significant ripple effect across the educational ecosystem.

For campuses looking to emulate the SAICM, the model of practice involves establishing a similar cycle of engagement that begins with creating awareness, facilitating active participation through volunteering, enhancing understanding via structured learning opportunities, and enabling leadership through mentorship. Key steps would include:

- Initiating Awareness Campaigns: Introduce the importance of academic integrity through interactive and compelling events to capture the campus community's interest.
- Developing Student Committees: Form student-led bodies to provide a structured platform for ongoing integrity initiatives.

- Implementing Learning Programs: Offer workshops and training focused on academic integrity and research ethics to build competence and understanding.
- Providing Practical Opportunities: Enable students to engage in research, policy formation, and organisational roles to apply their learning in real-world contexts.
- Recognizing and Celebrating Successes: Publicly acknowledge the efforts and achievements of student champions to motivate continued excellence.
- Cultivating Mentorship: Encourage experienced students to guide newcomers, ensuring knowledge transfer and the perpetuation of the integrity culture.

In conclusion, the SAIC Model is a culmination of years of tried and tested efforts that demonstrates the power of student-led initiatives in promoting academic integrity. It is a living model that goes beyond policy and enforcement, deeply integrating integrity into the fabric of campus life and extending its influence on the broader academic community. Institutions adopting this model of practice can look forward to not only strengthening their own culture of academic integrity, but also contributing to the global educational landscape's shift towards more ethical practices.

Acknowledgements We would like to thank ENAI WG Centre for Academic Integrity in the UAE (ECAIU) Student board and active committee for their support. We acknowledge Sheelagh Wallace, former Director of Registrars Student Services at University of Wollongong in Dubai and Akshita Bhatia, former student board member at ECAIU for their support in earlier stages of this work. We also acknowledge that ChatGPT has been used in certain parts of the chapter to enhance the sentence structure.

References

Ahshan, R. (2021). A framework of implementing strategies for active student engagement in remote/online teaching and learning during the COVID-19 pandemic. *Education in Science, 11*, 483. https://doi.org/10.3390/educsci11090483

Akbar, A., & Picard, M. (2019). Understanding plagiarism in Indonesia from the lens of plagiarism policy: Lessons for universities. *International Journal for Educational Integrity, 15*, 7. https://doi.org/10.1007/s40979-019-0044-2

Anohina-Naumeca, A., Birzniece, I., & Odiņeca, T. (2020). Students' awareness of the academic integrity policy at a Latvian university. *International Journal for Educational Integrity, 16*, 12. https://doi.org/10.1007/s40979-020-00064-4

Awad, M. K., Zogheib, B. & Alazemi, H. M. K. (2013). *A penalty scheme for academic dishonesty. Proceedings of IEEE International Conference on teaching, assessment and learning for engineering* (TALE) (pp. 580-584). IEEE Xplore. https://mohamadawad.com/blog/wp-content/uploads/2016/09/Tale2011.pdf

Benson, L., Rodier, K., Enström, R., & Bocatto, E. (2019). Developing a university-wide academic integrity E-learning tutorial: A Canadian case. *International Journal for Educational Integrity, 15*, 5. https://doi.org/10.1007/s40979-019-0045-1

Çelik, Ö., Yılmaz, T. S., Erçin, N., Razi, B. O., Akkan, S. C., Saka, I., & Uygun, E. (2022). Student involvement in upholding academic integrity: Student boards as praxis. In *Book of abstracts of the 8th European conference on academic integrity and plagiarism (ECAIP2022)* (pp. 230–232). ENAI.

Diego, L. A. B. (2017). Friends with benefits: Causes and effects of learners' cheating practices during examination. *IAFOR Journal of Education, 5*(2), 121–138. https://files.eric.ed.gov/fulltext/EJ1156266.pdf

Eury, J. L., & Treviño, L. K. (2019). Building a culture of honor and integrity in a business school. *Journal of Management Education, 43*(5), 484–508. https://doi.org/10.1177/1052562919850223

Farahat, A. (2022). Elements of academic integrity in a cross-cultural middle eastern educational system: Saudi Arabia, Egypt, and Jordan case study. *International Journal for Educational Integrity, 18*, 9. https://doi.org/10.1007/s40979-021-00095-5

Freedman, M. P. (2004). A tale of plagiarism and a new paradigm. *Phi Delta Kappan, 85*(7), 545–548. https://doi.org/10.1177/003172170408500714

Guerrero-Dib, J. G., Portales, L., & Heredia-Escorza, Y. (2000). Impact of academic integrity on workplace ethical behaviour. *International Journal for Educational Integrity, 16*, 2. https://doi.org/10.1007/s40979-020-0051-3

Hendershott, A., Drina, P., & Cross, M. (2000). Towards enhancing a culture of academic integrity. *NASPA Journal, 37*, 587–597. https://doi.org/10.2202/1949-6605.1119

Hussinger, K., & Pellens, M. (2017). *Guilt by association: How scientific misconduct harms prior collaborators*. ZEW discussion paper no. 17-051, Zentrum für Europäische Wirtschaftsforschung (ZEW). https://nbn-resolving.de/urn:nbn:de:bsz:180-madoc-436888

ICAI, International Centre for Academic Integrity. (2021). *Fundamental values of academic integrity*. https://academicintegrity.org/about/values

ICAI, International Centre for Academic Integrity. (2023). *Facts and statistics*. https://academicintegrity.org/resources/facts-and-statistics

John, S. R., Ramdas, S., Khan, S., Wilson, S., & Shabbir, R. (2021). *Formation of the student board at the UAE Centre for Academic Integrity–our initiatives and experience [paper presentation]*. Canadian Symposium on Academic Integrity. https://journalhosting.ucalgary.ca/index.php/ai/article/view/74169

John, S. R., Ramdas, S., Stumen, S. M. Wilson, S., & Khan, Z. R. (2023). *Role of student peer champions in building a culture of academic integrity*. Book of abstracts of the 9th European conference on ethics and integrity in academia (pp. 162-164). University of Derby: ENAI. https://academicintegrity.eu/conference/wp-content/uploads/2023/07/Book-of-Abstracts-Finalised-Version-for-PDF.pdf

Khan, Z. R. (2013). *Plagiarism-tracking software: Instructors' perception of software use vs. actual purpose of software. Proceedings of TARC international conference: Innovation and transformation in learning and teaching* (pp. 63-69), Tunku Abdul Rahman University College: Retrieved from https://ro.uow.edu.au/dubaipapers/554/

Khan, Z. R. (2014). *Developing a factor-model to understand the impact of factors on higher education students' likelihood to e-cheat* (doctoral dissertation). https://ro.uow.edu.au/theses/4545

Khan, Z. R. (2022). Lessons from delivering emergency online workshops that can enhance academic integrity values and skills among K-12 students. In *Book of abstracts of the 8th European conference on academic integrity and plagiarism (ECAIP2022)* (pp. 230–231). ENAI. https://academicintegrity.eu/conference/wp-content/files/2022/Book_of_Abstracts_2022.pdf

Khan, Z. R., Mumtaz, S., & Rakhman, S. S. (2019). Tracing the journey of two students' trajectory to becoming advocates of integrity—A case study. In *Proceedings from 5th international conference plagiarism across Europe and beyond and 3rd international conference shaping ethics in academia and society* (pp. 101–102). ENAI. https://academicintegrity.eu/conference/proceedings/2019/KhanMumtazSalma01.pdf

Khan, Z.R., Hemnani, P., Raheja, S. and Joshy, J. (2020a). Raising Awareness on Contract Cheating–Lessons Learned from Running Campus-Wide Campaigns. Journal of Academic Ethics, 18, 175–191. doi:https://doi.org/10.1007/s10805-020-09365-2.

Khan, Z. R., Thomas, T., & Sharma, V. (2020b). AI triage– Helping faltering students rehabilitate from academic misconducts. In Z. R. Khan, C. Hill, & F. Foltynek (Eds.), *Integrity in education for future happiness, 90–101* (pp. 90–101). Mendel University Press. https://academicintegrity.eu/conference/proceedings/2020/khan.pdf

Khan, Z. R., & Balasubrmanian, S. (2022). An ISM approach to modeling: Antecedents of E-cheating in higher education. In Management Association, I (Ed.), *Research anthology on interventions in student behavior and misconduct*. IGI Global. https://doi.org/10.4018/978-1-6684-6315-4.ch032

Khan, Z. R., Priya, J., & Tuffnell, C. (2022a). Culture of integrity—Institutional response to integrity during COVID19. *International Journal for Educational Integrity., 18*, 27. https://doi.org/10.1007/s40979-022-00118-9

Khan, Z. R., Hysaj, A., John, S. R., & Khan, S. A. (2022b). Transitional module on academic integrity to help K-12 students in the UAE prepare for the next stage of education. In S. Bjelobaba, T. Foltynek, I. Glendinning, V. Krasnica, & D. H. Dlabolova (Eds.), *Academic integrity: Broadening practices, technologies, and the role of students* (pp. 263–287). Springer. https://doi.org/10.1007/978-3-031-16976-2_15

Khan, Z. R., & Wallace, S. (2022, October 27). *Students as champions towards campus-wide efforts in building a culture of integrity* [video]. YouTube. https://www.youtube.com/watch?v=HGWLmnFIARY

Kostoulas-Makrakis, N., & Makrakis, V. (2020). Developing student-driven learning activities to promote refugee quality education through the CARE methodology. *International Journal of Early Years Education, 28*(2), 176–188. https://doi.org/10.1080/09669760.2020.1765091

Lee, E., & Hannafin, M. J. (2016). A design framework for enhancing engagement in student-centred learning: Own it, learn it, and share it. *Educational Technology Research and Development, 64*(4), 707–734. https://doi.org/10.1007/s11423-020-09842-w

Leslie, H. J. (2020). Trifecta of student engagement: A framework for an online teaching professional development course for faculty in higher education. *Journal of Research in Innovative Teaching & Learning, 13*(2), 149–173. https://doi.org/10.1108/JRIT-10-2018-0024

McCabe, D., & Pavela, G. (2000). Some good news about academic integrity. *Change, 32*(5), 32–38. https://doi.org/10.1080/00091380009605738

Norris, M. (2019). University online cheating–how to mitigate the damage. *Research in Higher Education Journal, 37*, 1–20. https://eric.ed.gov/?id=EJ1233121

Packalen, K., & Rowbotham, K. (2022). Student insight on academic integrity. In S. E. Eaton & J. Christensen Hughes (Eds.), *Academic integrity in Canada* (Vol. 1, pp. 353–375). Springer. https://doi.org/10.1007/978-3-030-83255-1_

Parnther, C. (2020). Academic misconduct in higher education: A comprehensive review. *Journal of Higher Education Policy And Leadership Studies, 1*(1), 25–45. https://doi.org/10.29252/johepal.1.1.25

Pittaway, S. M. (2012). Student and staff engagement: Developing an engagement framework in a Faculty of Education. *Australian Journal of Teacher Education, 37*(4). https://doi.org/10.14221/ajte.2012v37n4.8

Stoesz, B. M., Eaton, S. E., Miron, J., & Thacker, E. J. (2019). Academic integrity and contract cheating policy analysis of colleges in Ontario, Canada. *International Journal for Educational Integrity, 15*, 4. https://doi.org/10.1007/s40979-019-0042-4

Vučković, D., Peković, S., Blečić, M., & Dokovic, R. (2020). Attitudes towards cheating behavior during assessing students'performance: Student and teacher perspectives. *International Journal for Educational Integrity, 16*, 13. https://doi.org/10.1007/s40979-020-00065-3

Weber-Wulff, D., Anohina-Naumeca, A., Bjelobaba, S., Folytenk, T., Guerrero-Dib, J., Popoola, O., Sigut, P., & Waddington, L. (2023). Testing of detection tools for AI-generated text. *International Journal for Educational Integrity, 19*, 26. https://doi.org/10.1007/s40979-023-00146-z

Zhao, L., Mao, H., Compton, B. J., Peng, J., Fu, G., Fang, F., Heyman, G. D., & Lee, K. (2022). Academic dishonesty and its relations to peer cheating and culture: A meta-analysis of the perceived peer cheating effect. *Educational Research Review, 36*, 100455. https://doi.org/10.1016/j.edurev.2022.100455

Zwagerman, S. (2008). The scarlet P: Plagiarism, panopticism, and the rhetoric of academic integrity. *College Composition and Communication, 59*(4), 676–710. https://doi.org/10.58680/ccc20086674

Chapter 4
The Role of Unambiguous Assessment Briefs in Minimising Misconduct: Recommendations for Improving Assessment Rubrics

Shiva Das Sivasubramaniam, Zeenath Reza Khan, Salim Razi, and Dimitrios Kouroutsaltis

Abstract Academic integrity to uphold intellectual honesty and transparency has become the focus in higher education. Currently, the HE sectors worldwide have taken additional measures to minimise the likelihood of academic misconduct cases. Most of these interventions are student focussed, aiming to develop an integrity culture amongst students. A few studies have explored ways to deter plagiarism (or reduce the chances to plagiarise) by effective assessment strategies. It is an accepted argument that authentic assessment methods coupled with clear assessment briefs will reduce academic misconduct. Despite the continuous monitoring and careful mentoring, there were some reported incidences of ambiguous assessment briefs contributing to student misconduct. It is still not clear whether the assessment and their briefs are effectively produced with clear instructions about what is expected from the students (a) regarding literature surveys, (b) using data from collaborative group work, (c) using internet resources.

This chapter aims to revisit examples of ambiguous assessment briefs from different institutions worldwide and their role in contributing to plagiarism and academic misconduct cases. It will highlight reported cases of how the following

S. D. Sivasubramaniam (✉)
University of Roehampton, London, UK
e-mail: s.sivasubramaniam@roehampton.ac.uk

Z. R. Khan
University of Wollongong in Dubai, Dubai, UAE
e-mail: ZeenathKhan@uowdubai.ac.ae

S. Razi
Çanakkale Onsekiz Mart University, Çanakkale, Türkiye
e-mail: salimrazi@comu.edu.tr

D. Kouroutsaltis
University of Dundee, Dundee, UK
e-mail: dkouroutsaltis001@dundee.ac.uk

M. Davis, C. Baptista (eds.), *Ethics and Integrity in Education (Practice)*, Ethics and Integrity in Educational Contexts 10,
https://doi.org/10.1007/978-3-031-95686-7_4

assessment modalities: (a) restricted time constraint exams, (b) open book online tests, (c) take home assessment and (d) assessment briefs with contradictory instructions have provided chances for cheating and affected the performance of students with Specific Learning Difficulties (SpLDs). The chapter will elaborate examples of academic misconduct, especially plagiarism, resulting merely from ambiguous assessment briefs and inappropriately transformed online assessments. It will also discuss ways to reduce ambiguity in assessment briefs.

Keywords Assessment · Academic integrity · Honesty · Learning · Reflections · Misconduct

Introduction

Academic integrity and plagiarism avoidance have always been the focus in higher education (HE) to uphold intellectual honesty and transparency. The focus on deterring plagiarism gained momentum in 1997 (at least) in countries that took part in the development, launch and use of the first text matching software that was co-developed by I-Paradigm, JISC (Joint Information Systems Committee) and Northumbria Learning (Badge & Scott, 2009; Duggan, 2003). This so-called "anti-plagiarism software market" is continuing to grow at an annual growth rate of 22.87% between 2022 and 2027 (based on estimates by a leading market research company Technavio, 2024). This estimated growth is due to several factors, such as policies or positive initiatives such as worldwide institutional measures to curb academic misconduct, supportive government initiatives, and rising standards of university education with the focus on integrity training to the students. Other reasons may be related to the situational and technological changes that need proactive/reactive measures to maintain academic integrity (Wrigley & Smith, 2021). These include an increase in online assessment (especially after the Covid-19 pandemic), proliferation of ghost-writing services, growth of Generative AI (Gen-AI) and its widespread appeal to the students to deploy them in their work. As for plagiarism prevention, there is a plethora of studies on using a variety of software tools that can be employed as teaching tools to minimise the likelihood of students engaging in academic misconduct, especially plagiarism (Hayden et al., 2021; Mendonça et al., 2021; Johnson, 2023; Perkins et al., 2020).

Apart from the use of text matching software, the HE sectors worldwide have taken additional measures to minimise the likelihood of academic misconduct cases amongst staff as well as students. Most of these interventions are student focussed, aiming to develop a culture within the student community to become ethically sound individuals (Darling-Hammond et al., 2020). These include pre-sessional online tutorials on academic integrity (Benson et al., 2019), development of student honour codes, developing integrity modules (Benson & Enstroem, 2023), progressive sanctioning (penalties) linked to training (Sefcik et al., 2020), continuous training in academic writing (Morris, 2018). There are also a few studies that have explored ways to deter plagiarism (or reduce the chances to plagiarise) by effective assessment strategies including feedback-led continuous assessment, reflective assessments (Khan et al., 2021; Razı, 2023; Sivasubramaniam, 2013, 2014). Interestingly, questions remain about whether authentic assessments themselves can deter the chances of student plagiarism or not. In fact, Hughes and McCabe (2006) argue that authentic assessment methods coupled with clear assessment briefs will reduce academic misconduct. The term 'assessment brief' refers to a document that outlines the requirements, guidelines, and criteria for a specific assessment task. It serves as a comprehensive guide to help students understand what is expected of them, how they will be evaluated, and any specific details they need to consider while completing the assessment.

Others have suggested developing '*un-googlable*' questions especially for online assessment (Adama et al., 2023). One approach is to re-imagine the ways we assess student learning by understanding how students think, learn, and understand the instructions/assessment briefs. This includes (James et al., 2002):

1. Designing assessments for reflective application (as opposed to knowledge reproduction);
2. Providing unambiguous assignment requirements;
3. Producing authentic assessments (that are not recycled year-on-year).

Despite the continuous monitoring and careful mentoring by quality managers, there were some reported incidences of ambiguous assessment briefs contributing to student misconduct. This issue was further aggravated during the Covid-19 pandemic, when academics tried to transform traditional invigilated face-to-face assessments to unsupervised online tests (Holden et al., 2021; Khan et al., 2021). The academic authors of the present study (Authors 1 to 3), as members of the ethics and integrity advisory working group (EIAWG) of European Network for Academic Integrity (ENAI), in their capacity as educational advisors in different institutions, have dealt with several misconduct cases resulting from poor instructional guidelines (explained below). It is still not clear whether the assessment and their briefs are effectively produced with clear instructions about what is expected from the students (a) regarding literature surveys, (b) using data from collaborative group work, (c) using internet resources. This chapter aims to revisit examples of ambiguous assessment briefs from different institutions worldwide and their role in contributing to plagiarism and academic misconduct cases. Additionally, it will elaborate some student feedback about the difficulties they face with ambiguous assessment briefs. This feedback was obtained by the student author of this chapter (Mr Kouroutsaltis—DM) via peer conversations and openly available social media network searches. As the national student union representative (of UK), DM has been working closely with all UG and PG cohorts to gain their feedback relating to academic and extracurricular activities from 2020 till date. His input was used to enhance student support services by different universities. Considering these feedback concerns, the lead academic author has suggested some authentic assessment strategies with some examples from different subject areas.

Literature Survey on the Purpose of Assessment: Historical Perspectives

Assessments, in various forms, have been used to enhance the quality of learning experiences. They have a multitude of purposes, including educational [both as instructional (formative) and for evaluation (summative) of learning], vocational (to evaluate skill applications), psychological (to understand behaviours), and even political (to evaluate system performance) ends (Brown, 2022). However, educational assessment mainly focuses on measuring the students' 'understanding' of prescribed curriculum and/or to evaluate their abilities to apply skills, knowledge and performance. Assessment practices can be traced back to ancient civilizations such as Egypt, Greece, Rome, and India (De Young, 2019; Doyle, 1974; Morgan, 2021; NCERT, 2022). These civilizations employed various methods to assess students' abilities/knowledge, particularly in areas such as writing, arithmetic, and physical prowess. These assessment methods were employed to impart individuals' knowledge and skills, focussing on morals/ethics, integrity, individuality, and self-reliance.

One of the earliest and most well-known examples of formalised assessments is the Chinese imperial examination system, which originated during the Sui dynasty (sixth century AD) and continued until the early twentieth century (Feng, 1995). This system was used to select candidates for government positions based on their knowledge of Confucian classics and literary skills. In mediaeval Europe, assessments were primarily used within religious and academic institutions (Gillard, 2018). Monasteries and universities administered tests to assess students' understanding of theology, philosophy, and other subjects. These assessments played a crucial role in determining individuals' eligibility for clerical and scholarly positions. A significant change in the assessment practices came to light during the industrial revolution, when the concept of 'standardised tests' was introduced. The fact that these tests can be used (and re-used) for all students, and graded in the same manner for everyone, meant they were effective to evaluate large groups of individuals quickly and efficiently, for instance, IQ tests used in industrial and military recruitment developed (Richardson, 2022; Richardson & Norgate, 2015). In fact, the twentieth century witnessed significant advancements in educational assessment, particularly with the widespread adoption of standardised achievement tests (Brown, 2022). These tests were designed to measure students' academic proficiency in various subjects and became a central component of educational systems around the world. However, these tests only evaluate 'how much a person knows about a particular subject' (knowledge reproduction) rather than how effectively this knowledge can be used by that person (knowledge application in real life). It links a student's score as the sole judgement of that student's ability. In this way, it under-represents industrious students with higher levels of thinking and extraordinary aptitude for skill application. As a result of this, these students may be placed in remedial learning programs due to weak exam performance skills. Considering these issues, educational psychologists such as Benjamin Bloom tried to link the assessment with the cognitive skills of the learners (or their learning styles—Adams, 2015). Here the focus of assessments revolves around the level of study (See Fig. 4.1).

There has been a notable evolution in assessment styles in the twenty-first century. There has been a shift towards more dynamic, personalised, and technology-enhanced assessment approaches that aim to better measure individuals' skills, knowledge, and competencies. This change is mainly driven by advancements in technology, changes in educational paradigms, and shifts in societal needs (Twist, 2021). It is also undeniable that the Covid-19 pandemic played a big part in the introduction of new digital assessments and/or transformation of traditional assessment methods (such as invigilated examinations) to e-proctored assessment. This shift has, on the one hand, facilitated greater flexibility in test administration, quicker scoring, and the ability to incorporate multimedia elements (such as videos and interactive simulations) into assessments, but created challenges to maintain integrity on the other. Additionally, advances in data analytics have enabled more sophisticated analysis of assessment results, leading to insights that inform decision-making in various domains. Computer-based testing platforms offer greater flexibility, accessibility, and efficiency in administering assessments. A few examples of new and emerging (or transformed) methods of assessment are summarised in Table 4.1, some of which can be adopted as part of the solution for unethical use of GenAI in assessments (QAA, 2024; Lodge et al., 2023; Newman & Smith, 2016).

However, it should be noted, whilst these assessment strategies can be considered as innovative, their adoption into practice should depend on subject specialism, students' understanding, practical usage and above all, time constraints. For example, feedback-led continuous assessment is an excellent way to teach students via assessment, as it is structured in layers (or levels), with each layer building upon the

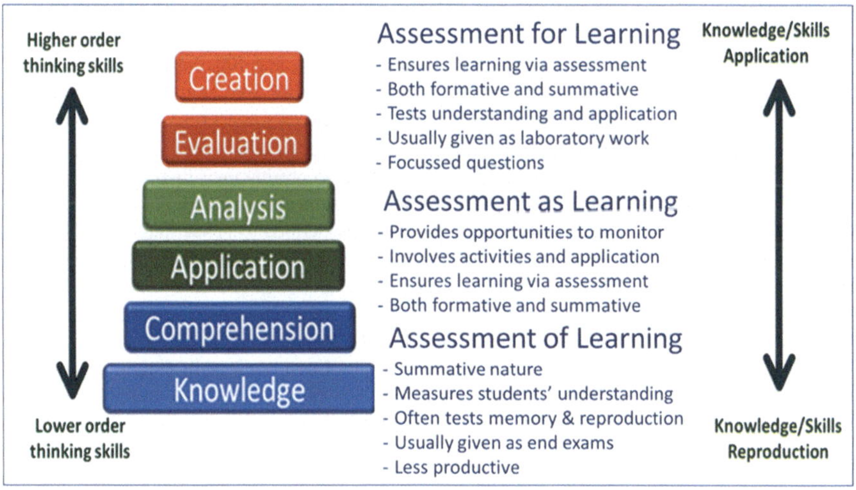

Fig. 4.1 The relationship between different orders of assessment with Bloom's Taxonomy (Bloom et al., 1956)

Table 4.1 Examples of innovative assessment methods of recent years

Assessment type and uses	Description
Adaptive testing (Stone & Davey, 2011) Mainly summative but can be used as formative	It dynamically adjusts the difficulty of test items based on the students' responses during the test. Unlike traditional tests where all test-takers receive the same set of questions, adaptive testing customises the test experience for everyone. It provides personalised assessment. This approach can lead to shorter test-taking times and more accurate measurement of proficiency levels.
Performance-based assessments (Shanteau et al., 2002) Mainly summative	It requires students to demonstrate their skills and knowledge in real-world contexts. These assessments include simulations, spot tests, projects, portfolios, and hands-on tasks. It provides an authentic measure of individuals' abilities. It has been used in applied science fields such as healthcare, engineering, and information technology in addition to the assessment of productive skills in language proficiency tests.
Formative assessment (also known as educative assessment; Morris et al., 2021; Stern, 2008; Wiggins, 1998) Formative (informal)	Formative assessment provides ongoing feedback to students to support their learning and development. Technology-enabled formative assessment tools, such as learning management systems and educational apps, have facilitated the implementation of these practices in classrooms.
Gamified assessments (Khan et al., 2021) Formative and summative	Gamified assessments make more engaging and motivating students. It incorporates elements of game design, such as points, levels, badges, and leaderboards, to incentivize participation and enhance user experience. This approach is particularly common in educational settings and employee training programs.
Feedback led continuous assessment. (Sivasubramaniam, 2014) Formative and summative	It promotes a more dynamic and student-centred approach to learning, where the focus is not just on grades, but on the learning process. It supports students by providing constructive feedback to help them understand their strengths and weaknesses, identify areas for improvement, and guide their learning journey.
Remote online proctoring (Akaaboune et al., 2022) Summative	Online proctoring is the technological equivalent of invigilated examination. It uses webcam and screen-sharing technologies to monitor students remotely and deter cheating during online exams. Although it has become increasingly common in distance education, it has been controversial due to privacy concerns.
Conference as assessment (Lund, 2014) Summative	It is organised in the same way as any other national conference, but it provides a space for students to present their final projects (either in the form of 15-min presentation or poster with defence). The presentation is assessed and moderated by academics.
Oral/digital oral examination (online/viva voce) (Scott & Unsworth, 2018)	This assessment is well suited for the evaluation of reflective and critical-thinking competencies alongside problem-solving and analytical abilities. It has been in use in medical related fields to test students' professional life skills including clinical-reasoning, decision-making and requisite. However, it is difficult to structure the process and often creates anxiety amongst the students.

previous one (Sivasubramaniam, 2014). (See section the role of assessment in learning below.) However, implementation of continuous assessment requires significant time and resources from both academics and students, particularly in large classes (Bjælde et al., 2022). Additionally, constant, and continuous assessment can create a high-pressure environment where students may feel overwhelmed by the need to perform well on multiple assessments (McKie, 2020; Popkova, 2018). This pressure can lead to stress and anxiety, detracting from the overall learning experience. It would shift the focus away from deep learning and students may prioritise short-term performance over long-term understanding. It may even lead the students to cheat as they focus on strategies to maximise their grades (Hernandez, 2012).

Despite these potential disadvantages, continuous assessment does promote active learning with student engagement, and most importantly supports ongoing improvement. Likewise, performance-based assessment is an excellent tool for testing students "contextual skills applications" (Heydarnejad et al., 2022). It enhances the students' knowledge and skills they have gained during their studies to their intended professions (or other related aspects of life). However, they often focus on specific skills or competencies, and thereby it may not capture other important aspects of learning (Playfoot et al., 2022). Ensuring consistency and fairness in performance-based assessments can be challenging, especially when assessing complex skills or tasks that may have multiple valid approaches or solutions (National Centre for Education Statistics, 1996). Students may experience increased pressure and anxiety during performance-based assessments, particularly if they are being observed or evaluated in real-time (Forstall, 2019). Above all, providing meaningful feedback on performance-based assessments can be challenging, particularly if assessments are conducted in real-time. This limited feedback may hinder students' ability to understand their strengths and areas for improvement. Therefore, careful consideration of the above said drawbacks can help its effective implementation by mitigating their impact on the learning experience.

The Role of Assessment in Learning

As stated above, assessments are integral components of the learning process. They provide valuable information and feedback that supports student growth, informs instructional practices, and ensures the attainment of learning goals and standards. In 2017, Archer highlighted the use of assessment in HE can be differentiated into (a) assessment to support learning; (b) assessment for accountability; (c) assessment for certification, progress, and attainment (career) (Archer, 2017). She further linked these three different types as an "assessment purpose triangle". Unfortunately, nowadays, many HE organisations are mainly focusing on achievement-based attainments. In fact, the introduction of quality-based university classifications such as the Teaching Excellence Framework (TEF) of the UK, university league tables, and university world ranking have resulted in universities focusing only on attainments such as grades, test scores, and degree classifications. It is an undeniable fact that attainment is an important aspect of university education. However, an exclusive focus on this metric alone can have significant drawbacks on the learning process of students. There has been a trend in institutions reconsidering assessment regulations and practices to enhance student success and increase pass rates (Elkington, 2015; Office for Students, 2022; Weatherton & Schussler, 2021; Orr & Conaway, 2016). Some of which has resulted in cause for concern by professional, statutory, and regulatory bodies (PSRBs) redefining their specifications of a "pass mark" for skills-oriented modules. For example, in 2024, the Institute of Biomedical Science (IBMS) changed its "Criteria and Requirements for the Accreditation of BSc (Hons) Degrees in Biomedical Science" as follows (Institute of Biomedical Science, 2024):

> Students must achieve a pass mark of 40% **in all assessment components** for modules that cover the clinical laboratory sciences subject areas. Condonement/compensation will not be permitted for these modules or for any other module(s) that contribute significantly to the benchmark statement and have learning outcomes that students achieve that cannot be evidenced elsewhere.

As a fellow of IBMS, the lead author of this article understands that this was a direct reflection of some university's initiatives to introduce an "cumulative pass" (by which students could pass if they achieved an overall mark of 35% as a measure of achieving the learning outcomes of that module). Assessment regulations may indeed result in higher pass rates, particularly if they address underlying factors contributing to student attrition or academic underachievement. Therefore, it is important for universities to approach such revisions thoughtfully and ethically.

The primary focus should be on fostering meaningful learning experiences, promoting academic integrity, and equitably supporting all students in achieving their educational goals. At this juncture, it is important for the academics (and HE organisations) to focus on the following questions before setting up an assessment task.

- What is the purpose of the assessment?
- What specific knowledge will be gained?
- What are the future applications needed for this assessment?
- How might this assessment be useful to improve the learning experience?

Assessments can be used in different contexts within education and higher education. They can be used to (a) test the learning (assessment of learning), (b) teach or make the student learn (assessment for learning) and (c) initiate self-directed learning (assessment as learning) (NSW Education Standards Authority, n.d.; Rowe, n.d.). Figure 4.1 depicts the authors' attempt to extrapolate these in relation to Bloom's taxonomy of students' cognitive skills ranging from lower-order (that require less cognitive processing) to higher-order skills (that require deeper learning, with a greater degree of cognitive processing and critical analysis).

It is essential to promote meaningful learning experiences, fostering critical thinking skills, reducing bias, and accurately measuring student learning outcomes. The modes of assessments in HE often range from assessment of learning at lower levels (such as NFQ 3 or 4), to assessments for learning at degree/master's level (NQF 5 to 7). As stated in the Advance HE directives on enhancing student success, the modes of assessment at all levels should mainly focus on empowering and engaging students in relation to subject specific competencies through assessment design (Advance, n.d.).

How can students' learning be enhanced through assessment? Unambiguous assessment briefs with clear instructions on expectations play an important part in student engagement with the assessment, by minimising the chances of academic misconduct cases. Clear assessment briefs can promote fairness, transparency, and effectiveness in the assessment process. When expectations are clearly articulated, students are better equipped to seek clarification when needed, leading to smoother interactions and a more positive learning experience. It also ensures accurate evaluation of student performance. Authors and other academics have noticed that students are more likely to cheat when assessment briefs are not clear, as students may feel uncertain about what is expected of them or how they will be evaluated (see next section). This uncertainty can create conditions that increase the likelihood of cheating for several reasons.

Ambiguous Assessment Briefs Leading to Academic Misconduct

Clarity in assessment instructions is essential for promoting academic integrity and reducing the likelihood of cheating among students. Unclear and/or ambiguous assessment briefs can create an environment ripe for academic dishonesty by fostering anxiety, perceived unfairness, and opportunities for exploitation. The ambiguous assessment briefs may sometimes result from simple mistakes such as improper recycling of questions. Our conversations with student representatives (via DM) have highlighted several examples of simple mistakes within the assessment briefs. These include, (a) lack of clarity—some assessment briefs were written (or used) general rubrics instead of detailing specific expectations, which has led to uncertainty over what is expected to meet the marking criteria; and (b) subject to misinterpretation—misleading information on what is expected increases the chances of misinterpretation of what is being asked and what is required to get the required marks. They pointed out that an ambiguous assessment brief would also impact their learning engagement.

According to them "*if the assessment brief is itself unclear, we would focus on interpreting the assignment brief rather than the actual assessment*" (Student A). "*Overall if I find an assessment brief is ambiguous it leads to wasted time trying to work out the requirements which leads to me feeling frustrated and demotivated, then causes anxiety whilst awaiting the grades and feedback*" (Student B). One student representative from a UK institution has shared some examples of such mistakes. He stated that "*one brief appeared to have been recycled from the previous year and not updated, and so the key dates were incorrect. Furthermore, the required format for the file name of your submission did not match with the format that was requested in the Turnitin submission*" (Student C). On a closer look, the brief itself was written in such a way that expected students to carry out multiple tasks outside the question. This has made the students overwhelmed, especially in the final year, resulting in distress. He added "*some briefs provide clear guidance on what is expected to be included and the word count for each section of the work, while others provide little on how to structure your work*". Although this could be beneficial in terms of developing independence, it would often make it difficult to understand the expectations of the module leader. He also showed an assessment brief which simply stated that the content of the coursework would be introduced during one of the revisions sessions, which never happened.

There were no marking rubrics, and in fact it simply lacked information for an assessment brief. Other forms of simple mistakes include the assessment brief expecting two different referencing styles. This has created confusion and affected students' confidence when writing this piece of coursework. When students tried to clarify with the lecturer, their email was replied after the submission deadline. They further added "*other students were experiencing the same issue*". Another student from the Middle Eastern university pointed out "*In my experience as a student, a badly written assessment brief can be a major roadblock in my academic career*" (Student D). This student recalled an incidence in which she was unable to demonstrate her expertise to the maximum degree mainly due to the unclear assessment brief. They further added "*It is difficult for me to modify my work to fulfil the requirements of the module since there are no explicit instructions regarding the evaluation criteria and learning objectives. I often spend too much time second-guessing myself and trying to read the instructions*". Overall, it was clear that even simple mistakes in the assessment briefs with wrong submission dates, unclear referencing expectations can be challenging for the students to plan and prioritise their work effectively. Instead of focusing on the actual content of the assignment, they often spend too much time trying to read the instructions. This uncertainty can be frustrating resulting in anxiety.

Considering the above student responses, we further investigated the effects (and examples of) ambiguous assessment briefs resulting in unintentional academic misconducts. Plagiarism is more prevalent in some forms of assessment, especially in unsupervised knowledge reproducing evaluations which give the temptation to 'Google' answers. These types of questions may be unavoidable in lower National Qualification Framework (NQF) levels, such as pre-university foundation degrees, and first year of entry to higher education. It is appropriate for these assessments to be offered as invigilated/supervised tests. During the pandemic, institutions were forced to offer these assessments online. Here we would like to highlight reported cases of how the so-called "restricted time constraint exams" have provided (a) opportunities for cheating and (b) affected the performance of students with Specific Learning Difficulties (SpLDs). Timed assessments often do not accurately reflect the true abilities and knowledge of students with SpLDs. These students might understand the material well but require more time to process and express their thoughts. As a result, timed tests can provide an inaccurate measure of their actual academic capabilities which can significantly impact students with SpLDs in various ways. SpLDs often have challenges that make it difficult for them to perform well under timed conditions, and this can affect their academic performance and overall well-being. Students with SpLDs, such as dyslexia, ADHD, or processing speed deficits, may experience heightened anxiety and stress during timed assessments (Blanchet & Assaiante, 2022). The pressure to complete tasks within a limited timeframe can exacerbate their difficulties and hinder their ability to concentrate and recall information. Although this type of time constrained examinations/assessments have been in use, especially in distant learning programmes, the Covid-19 pandemic has forced the universities to adopt this. In these online multiple-choice or short answer tests, questions were linked to a specific time (usually 3 seconds), which was clearly stated in the assessment briefs. It was thought this would minimise the time needed to 'Google' the answers. However, some students obtained help from their friends/family with the searches.

Although it is difficult to prove this is a form of exam cheating, many universities reported an increase in high performance in MCQs, especially during Covid-19 pandemic. On the other hand, at least in some countries, students with special needs/documented disabilities were not able to complete the task in time (Barnard-Brak et al., 2010). In fact, a critical analysis report published in 2020 has clearly highlighted the main reasons against using time constrained assessment (Gernsbacher et al., 2020), (a) it limits the students' thinking time, as estimated time does not consider the "slow thinkers" and therefore does not reflect student knowledge and (b) it is impossible to gauge additional test-taking time for those with documented disabilities and the international students for whom English is a second language or older students with learning difficulties.

As for ambiguous assessment briefs, we have noticed several reported cases in different institutions worldwide, especially during (and immediate post-Covid) periods. Online open book tests have become very popular after the pandemic. On advanced NQF levels such as second, and final years of undergraduate studies (with a lesser extent of level 7—postgraduates), some institutions (within UK and beyond) offered a slightly modified version of 'open book online tests' where the assessment briefs stated, "answers/information obtained from other sources should be properly acknowledged" and "Any unacknowledged information will be considered as academic misconduct" (source anonymised). As a result, there were incidents where students cut and pasted small paragraphs and gave proper attributions and references. These students claimed that they had followed the assessment brief. In this case, the questions should have been intellectually challenging for the students to apply their knowledge/skills. Furthermore, there should have been a clear definition about the limitations of the "open book test" offered. The accepted definition of an open book test "allows referring to the notes and consulting reference/course materials" (as defined by Merriam-Webster, 2023 online dictionary), students' claims were deemed to be acceptable. Ambiguous instructions like this led students to misinterpret the requirements of the assessment, resulting in unintentional errors, and misconduct, (such as plagiarism or cheating) as they are unsure about what is expected of them and resort to inappropriate means to complete the assignment.

In addition to the examples stated in Table 4.2, it is worth pointing out the long-standing concern that has persisted in higher education assessments regarding assessment of group work, as it lacks clear guidelines on individual versus collective responsibilities, creating potential confusion about grading criteria and group dynamics. The instructions state, "*you are expected to work with another person to assimilate data from online sources, however a percentage of marks would be allocated for your own initiative*!" without any instructions/clarifications on (a) how would their 'paired work' be separated from individual initiative; (b) the percentage

Table 4.2 Example ambiguous assessment briefs that led to student misconduct/complaints

Ambiguous assessment briefs	Issue(s) highlighted by the students (or lodged as a complaint) in the programmes committee meeting
'Create a multimedia presentation about black box algorithms in medicine'. (for black box algorithms—please refer Durán & Jongsma, 2020.	No clear indication on the student is expected to write the uses of black box algorithms, advantages or disadvantages.
'Four of you are expected to work as a group to discuss the effects of electric cars on environmental issues.' However, you are expected to complete the assignment individually.	It is not specified how much collaboration is allowed or how individual contributions should be acknowledged. Some students might collaborate extensively without citing each other's contributions, leading to accusations of plagiarism.
'You can use GenAI for obtaining initial information, as long as they are properly acknowledged.	Students complained that they were not explained anything about using gen-AI or how to acknowledge it.
'Analyse the provided case study about this patient and discuss its implications'.	The question does not specify what is expected, whether a diagnosis with justification? Or how this patient is going to deteriorate or prognose?
'You are allowed to use GenAI in this report; however, you are advised to read the university policy found in—Link (the link is anonymised)'.	Student representatives complained that the university policy only stated "*individual instructors decide their own policy*" which led to confusion.
'Using your preferred online based software or other visual programming tool or software development kit, you are expected to source code for a simple computer game'.	One student who was accused of plagiarising the code claimed that she thought the instructor only expected source code for a game and the fact that she was allowed to use any online tools, she just copied.
'Evaluate the effectiveness of the UK healthcare system reforms discussed in the textbook'.	No clear information about the textbook. The instruction does not specify the criteria for evaluating effectiveness or the scope of the analysis.
'Critically evaluate the phrase "one mRNA code only produces one amino acid, but one amino acid can be produced by more than one code". You are expected to provide examples.	The phrase itself was confusing, urging the student to consult online resources which simply provide the exact critical analysis with examples (indirectly) urging the students to copy and complete the work.

(continued)

Table 4.2 (continued)

Ambiguous assessment briefs	Issue(s) highlighted by the students (or lodged as a complaint) in the programmes committee meeting
'You are not allowed to use any internet sources to complete this exercise' you are expected to use the notes, and the directives discussed in the laboratory class to complete this assignment.' 'Any internet sources used in this exercise should be properly acknowledged.'	Confusing and conflicting instructions. Students were not sure whether to use the internet source or not. This issue was raised as a complaint against the lecturer and the university.
'Write a critical review on Balahmar, et al. 2021 by summarising the main points of the article in your own words."	No specifications were given on how much of the text can be directly paraphrased or what level of transformation is expected. Some students inadvertently relied upon original wording, leading to unintentional plagiarism.
This is a formative assessment only. You are expected to use as many materials as possible to complete this work. Submissions will be checked for plagiarism.	It is not specified whether students can include content from lecture notes, supplementary readings, or online sources without proper attribution. Some students might incorporate unattributed material from these sources, leading to plagiarism.

allocated for individuality and (c) what are the preferred 'online sources'. This has resulted in students submitting almost identical critical analysis. Other examples of ambiguous (or not clearly defined) assessment briefs and rubrics are given below in Table 4.2 (these examples were gathered (a) by the student author and (b) from programme committee-based student concerns and complaints mainly from the UK, Europe and Middle-Eastern countries.

The examples given above clearly show that the assessment briefs should be better scrutinised. They highlight the fact that the instructions should provide clear guidance to complete the task, define expectations, and offer guidance on how students should approach the tasks while still allowing them to utilise their resources effectively. When assessment briefs lack clarity regarding expectations, students may be more prone to engaging in misconduct due to confusion about what constitutes acceptable behaviour.

Measures to Minimise Ambiguous Assessment Briefs

It is an accepted fact that minimising ambiguous assessment briefs is essential for ensuring fairness and clarity in academic evaluation and enhancing effective learning outcomes for students. A good assessment brief will also promote constructive dialogue between staff and students, narrowing the gap between staff expectations and student performance. As members of the ethics and integrity working group with ENAI, the first three authors, analysed the issues and discussed with other

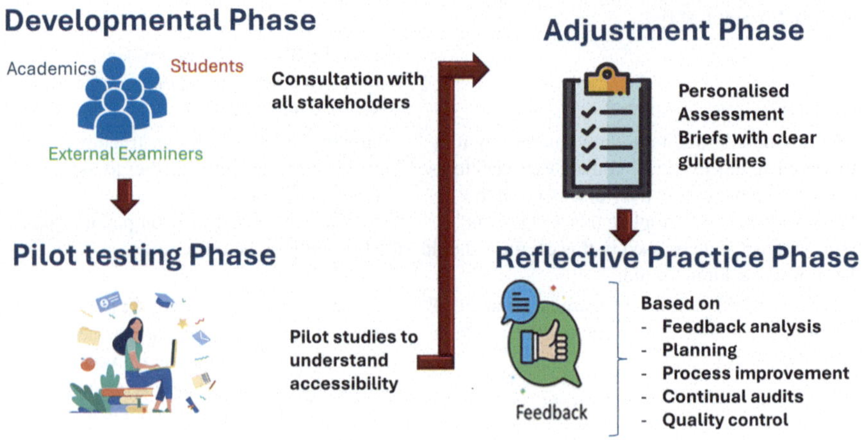

Fig. 4.2 Suggestive guidelines linking with different phases of assessment development to ensure unambiguous assessment briefs/rubrics

educators within the integrity sciences to provide the following sequence of guidelines for minimising ambiguous assessment briefs (see also Fig. 4.2). The figure depicts different phases of assessment development linking them to suggestive guidelines to ensure unambiguous assessment briefs/rubrics.

1. **Consultation:** Consulting with colleagues (and students) during the design phase of assessment tasks is essential to identify and address any potential ambiguities. This process in some universities is undertaken during the programme development stage. However, this should be comprehensive and carried out periodically to address any ongoing (during programme/module restructuring) or unexpected (such as pandemic) changes. Additional consultation using external examiners would also be useful.
2. **Creating personalised rubrics:** Assessment rubrics should be linked to the learning outcomes of the module/subject. Developing and distributing rubrics with detailed guidelines outlining the expectations, criteria for evaluation, and the purpose of the assessment is essential. This clarity helps students understand what is expected of them. Especially in course work type of summative assessment, students should be able to understand what is expected. In other words, rubrics should have information on how different grade bands are allocated (such as what they should do to gain a pass grade, a third class, second class upper & lower divisions or the first-class mark). This would not only provide an understanding on what is expected for each of the different grade bands; but also minimise any student appeals against the marks. Rubrics make grading more transparent and consistent (i.e. using generic rubrics should be avoided). It is also worth offering examples of both good and poor-quality work to illustrate the standards expected. Concrete examples help students grasp the level of quality required.

3. **Consideration for students with SpLDs.**

   Assessment should consider the students with SpLDs by providing additional time for students to reduce anxiety and allow them to demonstrate their true abilities.

   Allowing breaks, providing a quiet testing area, or offering exams in multiple shorter sessions can help students manage their time more effectively. Where possible, utilising untimed or differently structured assessments, (such as projects, oral presentations, or take-home exams), can provide a more accurate measure of a student's understanding. Most importantly, educators should be trained to recognize and support the needs of students with SpLDs, ensuring they can provide appropriate accommodations and modifications to assessments.
4. **Pilot testing:** Pilot testing new assessment tasks before implementing them widely is essential to identify any ambiguities or issues and make necessary revisions. It is a crucial step in the assessment development process, allowing academics to refine and validate new tasks before full-scale implementation. It ensures that the assessment accurately measures the intended constructs, is unbiased, and provides meaningful insights for decision-making purposes. This helps ensure that the tasks are appropriately challenging for the intended audience. Tasks that are too easy may not accurately measure the participants' abilities, while tasks that are too difficult may lead to frustration and disengagement. The feedback obtained from other academics, or the students can offer valuable insights into participants' experiences and highlight any issues that may not have been apparent during the design phase. By testing the tasks before widespread implementation, academics can make sure the assessment briefs are validated and reliable tasks with accessibility to all students, including those with disabilities or language barriers. This can increase the students' trust in the assessment process and the decisions based on the results.
5. **Internal/External moderation (scrutinising assessment):** Having assessment questions checked by other academic(s) strengthens the integrity of the assessment process and helps ensure the questions are unambiguous. What is clear to one academic may not be clear to another academic or students. Therefore, it is essential for the assessments (and their rubrics) to be checked by both internal and external moderators. This has become a common practice in many universities. By this pre-assessment moderation process, institutions can ensure the quality of the questions by obtaining a fresh perspective from another academic who could also identify any ambiguities, inconsistencies, or errors that the original author might have overlooked. This helps maintain the integrity, validity and reliability of the assessment and ensures that the questions accurately measure the intended constructs (or the learning outcomes), and they are understandable to a diverse audience. This is especially important in educational settings where students come from various cultural and linguistic backgrounds. Most importantly, it ensures consistency in assessment across different modules or courses by aligning assessment criteria with learning outcomes and program objectives.

6. **Providing clear feedback with feed-forward**: It is also important to encourage students to seek clarification if they find any part of the assessment brief ambiguous.

   Additionally, providing timely feedback on drafts or preliminary work to guide students in the right direction will ensure reflective learning. Incorporate peer review processes where students provide feedback on each other's work based on the assessment criteria. This not only helps clarify expectations but also fosters a collaborative learning environment.
7. **Reflective practice**: Regularly reflecting on past assessments to identify areas for improvement is essential to refine future assessment briefs accordingly.

In this way, the institutions can ensure a strong alignment between the intended learning outcomes and the assessed tasks with clear instructions to ensure that assessments accurately measure the desired learning objectives.

Assessment rubrics need to be linked with module learning outcomes during the initial programme development phase with wider consultation with important stakeholders (i.e. consultative phase should be linked to programme development). Assessments are then piloted with students (as mock tests) to obtain student feedback on clarity, inclusivity. Based on the feedback, assessments are readjusted to make them unambiguous. Finally, the reflective phase makes sure the ongoing year-on-year assessments contain unambiguous instructions/rubrics.

Conclusion

This chapter has highlighted some examples of ambiguous assessment briefs and how they may contribute to academic misconduct cases amongst the students. Persistent ambiguity in assessment briefs may even result in a negative perception of education among students. When students feel that assessment criteria are unclear or arbitrary, they may question the value and fairness of their educational experiences, leading to disengagement and disillusionment. A well-organised, educational, and student-focused assessment briefs that set clear expectations, encourage critical thinking, and create a stronger bond with the material should be given top priority by educational institutions and teachers to address these issues. Clear guidelines help students understand expectations, reducing the likelihood of inadvertent plagiarism or other forms of misconduct. When students are fully aware of the requirements and the standards by which they will be judged, they are more likely to produce original work. Additionally, clarity in assessment briefs helps reduce confusion and anxiety, fostering an environment where academic integrity is prioritised. By providing precise instructions, educators can guide students toward proper research and citation practices, ultimately cultivating a culture of honesty and ethical scholarship. Considering the above, we recommend the following actions for academics to improve assessment briefs by providing clear instructions:

- **Grade criteria**: Ensure that rubrics clearly define criteria and expectations for each grade level. Each element of the rubric should be described in detail, leaving no room for ambiguity about what constitutes excellent, good, satisfactory, and poor performance.
- **Alignment with Learning Outcomes**: Rubrics should be closely aligned with the learning outcomes of the module and/or course. Each criterion should directly relate to a specific learning outcome, ensuring that assessments measure what they are intended to.
- **Exemplars to understand the expectations**: Provide students with examples of high-quality work that meets each criterion. This can help them better understand what is expected and how to achieve high marks.
- **Consistency and Fairness**: Ensure that rubrics are applied consistently across all students, by specifically focussing on SpLDs. By understanding and addressing the unique needs of students with special learning difficulties, educators can create a more equitable and supportive learning environment that allows all students to succeed. Regularly review and calibrate rubric standards among multiple graders to maintain fairness and objectivity.
- **Student Involvement**: Involve students in the creation or review of rubrics. This can enhance their understanding and buy-in, as they feel more connected to the assessment process.
- **Feedback/feed forward to enhance learning**: Use the rubric as a tool for providing detailed, constructive feedback. Highlight areas of strength and specific aspects that need improvement, guiding students on how to enhance their future performance.
- **Regular Reviews with reflections**: Periodically review and update rubrics to ensure they remain relevant and effective. Incorporate feedback from students and educators to continuously improve the assessment process.

We recommend institutions provide training for educators on how to develop and use rubrics effectively. This includes how to design clear, objective criteria and how to apply the rubric consistently. By implementing these recommendations, we believe, the educational institutions can enhance the effectiveness of their assessment rubrics, further reducing the incidence of academic misconduct and promoting a culture of integrity and excellence.

References

Adama, E. A., Graf, A., Adusei-Asante, K., & Afrifa-Yamoah, E. (2023). COVID-19 and alternative assessments in higher education: Implications for academic integrity among nursing and social science students. *International Journal for Educational Integrity, 19*(8). https://doi.org/10.1007/s40979-023-00129-0

Adams, N. E. (2015). Bloom's taxonomy of cognitive learning objectives. *Journal of the Medical Library Association, 103*(3), 152–153. https://doi.org/10.3163/1536-5050.103.3.010

Advance H. E. (n.d.). *Essential frameworks for enhancing student success*. https://www.advance-he.ac.uk/guidance/teaching-and-learning/essential-frameworks-enhancing-student-success

Akaaboune, O., Blix, L. H., Carrington, L. G., & Henderson, C. D. (2022). Accountability in distance learning: The effect of remote proctoring on performance in online accounting courses. *Journal of Emerging Technologies in Accounting, 19*(1), 121–131. https://doi.org/10.2308/JETA-2020-040

Archer, E. (2017). The assessment purpose triangle: Balancing the purposes of educational assessment. *Frontiers in Education: Assessment, Testing and Applied Measurement, 2*. https://doi.org/10.3389/feduc.2017.00041

Badge, J., & Scott, J. (2009, November). *Dealing with plagiarism in the digital age*. Higher Education Academy EvidenceNet. http://evidencenet.pbworks.com/w/page/19383480/Dealing%20with%20plagiarism%20in%20the%20digital%20age

Barnard-Brak, L., Lectenberger, D., & Lan, W. Y. (2010). Accommodation strategies of college students with disabilities. *The Qualitative Report, 15*, 411–429. https://doi.org/10.46743/2160-3715/2010.1158

Benson, L., & Enstroem, R. (2023). A model for preventing academic misconduct: Evidence from a large-scale intervention. *International Journal for Educational Integrity, 19*(25). https://doi.org/10.1007/s40979-023-00147-y

Benson, L., Rodier, K., Enstroem, R., & Bocatto, E. (2019). Developing a university-wide academic integrity e-learning tutorial: A Canadian case. *International Journal for Educational Integrity, 15*, 1–23. https://doi.org/10.1007/s40979-019-0045-1

Bjælde, O. E., Jørgensen, T. H., & Lindberg, A. B. (2022). Continuous assessment in higher education in Denmark. *Dansk Universitetspædagogisk Tidsskrift, 12*(23), 1–19. https://doi.org/10.7146/dut.v12i23.25634

Blanchet, M., & Assaiante, C. (2022). Specific learning disorder in children and adolescents, a scoping review on motor impairments and their potential impacts. *Children (Basel), 9*(6), 892. https://doi.org/10.3390/children9060892

Bloom, B. S., Engelhart, M. D., Furst, E. J., Hill, W. H., & Krathwohl, D. R. (1956). *Taxonomy of educational objectives: The classification of educational goals. Handbook I: Cognitive domain*. David McKay Company.

Brown, G. T. L. (2022). The past, present, and future of educational assessment: A transdisciplinary perspective. *Frontiers in Education, 7*. https://doi.org/10.3389/feduc.2022.1060633

Darling-Hammond, L., Flook, L., Cook-Harvey, C., Barron, B., & Osher, D. (2020). Implications for educational practice of the science of learning and development. *Applied Developmental Science, 24*(2), 97–140. https://doi.org/10.1080/10888691.2018.1537791

De Young, G. (2019). Diagrams in ancient Egyptian geometry: Survey and assessment. *Historia Mathematica, 36*(4), 321–373. https://doi.org/10.1016/j.hm.2009.02.004

Doyle, K. O. (1974). Theory and practice of ability testing in ancient Greece. *Journal of the History of Behavioral Sciences, 10*(2), 202–212. https://doi.org/10.1002/1520-6696(197404)10:2<202::AID-JHBS2300100208>3.0.CO;2-Q

Duggan, F. (2003). The plagiarism advisory service. *Library and Information Research, 27*(86), 37–42. http://www.lirg.org.uk/lir/pdf/article86d.pdf

Durán, J. M., & Jongsma, K. R. (2020). Who is afraid of black box algorithms? On the epistemological and ethical basis of trust in medical AI. *Journal of Medical Ethics, 47*(5), 329–335. https://doi.org/10.1136/medethics-2020-106820

Elkington, S. (2015). *Essential frameworks for enhancing student success: Transforming assessment in higher education*. A guide to the Advance HE Framework. Advanced HE. https://s3.eu-west-2.amazonaws.com/assets.creode.advancehe-document-manager/documents/advance-he/EFSS_%20Guide-Transforming%20Assessment_1589628649.pdf

Feng, Y. (1995). From the imperial examination to the national college entrance examination: The dynamics of political centralism in China's educational enterprise. *Journal of Contemporary China, 4*, 28–56. https://doi.org/10.1080/10670569508724213

Forstall, M. (2019, May 14). *The Classroom: The disadvantages of performance-based assessment* [Blog]. https://www.theclassroom.com/high-achiever-vs-gifted-student-13699.html

Gernsbacher, M. A., Soicher, R. N., & Becker-Blease, K. (2020). Four empirically based reasons not to administer time-limited tests. *Translational Issues in Psychological Science, 6*(2), 175–190. https://doi.org/10.1037/tps0000232

Gillard, D. (2018). *Education in the UK: A history*. History Docs Articles (HDA). www.education-uk.org/history

Hayden, K. A., Eaton, S. E., Pethrick, H., Crossman, K., Lenart, B. A., & Penaluna, L. A. (2021). A scoping review of text-matching software used for student academic integrity in higher education. *Education Research International, 2021*, 4834860. https://doi.org/10.1155/2021/4834860

Hernandez, R. (2012). Does continuous assessment in higher education support student learning? *Higher Education, 64*(4), 489–502. https://doi.org/10.1007/s10734-012-9506-7

Heydarnejad, T., Tagavipour, F., Patra, I., & Khafaga, A. F. (2022). The impacts of performance-based assessment on reading comprehension achievement, academic motivation, foreign language anxiety, and students' self-efficacy. *Language Testing in Asia, 12*(51). https://doi.org/10.1186/s40468-022-00202-4

Holden, O. L., Norris, M. E., & Kuhlmeier, V. A. (2021). Academic integrity in online assessment: A research review. *Frontiers in Education, 6*, 1–13. https://doi.org/10.3389/feduc.2021.639814

Hughes, J. M. C., & McCabe, D. L. (2006). Understanding academic misconduct. *Canadian Journal of Higher Education, 36*(1), 49–63. https://doi.org/10.47678/cjhe.v36i1.183525

Institute of Biomedical Science (IBMS). (2024, September). *Criteria and requirements for the accreditation of BSc (Hons) degrees in biomedical science.* https://www.ibms.org/resources/documents/bsc-initial-accreditation-guidance-for-cohorts-from-sept-2024/

James, R., McInnis, C., & Devlin, M. (2002). *Assessing learning in Australian universities: Ideas, strategies and resources for quality in student assessment.* Centre for the Study of Higher Education, The University of Melbourne and the Australian Universities Teaching Committee. http://www.cshe.unimelb.edu.au/assessinglearning/docs/AssessingLearning.pdf

Johnson, C. (2023). *Understanding academic integrity and plagiarism in the digital age: Can digital forensics techniques help prevent and detect academic misconduct? PhD Thesis.* University of South Wales/Prifysgol De Cymru. https://pure.southwales.ac.uk/ws/portalfiles/portal/15727327/Understanding_academic_integrity_and_plagiarism_in_the_digital_age_Clare_Johnson_librarycopy.pdf

Khan, Z. A., Draper, M., Bjelobaba, S., Razi, S., & Sivasubramaniam, S. D. (2021). Academic integrity outreach efforts–making education accessible and inclusive. In S. Bjelobaba, T. Foltýnek, I. Glendinning, V. Krásničan, & D. H. Dlabolová (Eds.), *Academic integrity: Broadening practices, technologies, and the role of students. Ethics and integrity in educational contexts* (Vol. 4, pp. 17–28). Springer. https://doi.org/10.1007/978-3-031-16976-2_1

Lodge, J. M., Howard, S., & Broadbent, J. (2023, May 1). Assessment redesign for generative AI: A taxonomy of options and their viability. *LinkedIn Article.* https://www.linkedin.com/pulse/assessment-redesign-generative-ai-taxonomy-options-viability-lodge/

Lund, N. (2014). Student conference as assessment. *Good practice exchange*, centre for excellence in learning and teaching [Youtube video]. https://www.youtube.com/watch?v=_Y3EK-9ORkQ

McKie, A. (2020). *Continuous assessment 'may be more stressful' for students.* Times Higher Education Blog. https://www.timeshighereducation.com/news/continuous-assessment-may-be-more-stressful-students

Mendonça, M., Martinelli, J. E., Sadauskas, D., & de Andrade, D. F. (2021). Plagiarism detection software: Help or hindrance? *Research Integrity and Peer Review, 6*(1), 1–13. https://doi.org/10.1186/s41073-021-00113-7

Merriam-Webster. (2023). *Online dictionary*. https://www.merriam-webster.com

Morgan, T. (2021). Assessment in Roman education. *Assessment in Education: Principles, Policy & Practice, 8*(1), 11–24. https://doi.org/10.1080/09695940120033225

Morris, E. J. (2018). Academic integrity matters: Five considerations for addressing contract cheating. *International Journal for Educational Integrity, 14*(15), 1–10. https://doi.org/10.1007/s40979-018-0038-9

Morris, R., Perry, T., & Wardle, L. (2021). Formative assessment and feedback for learning in higher education: A systematic review. *Review of Educational Research, 91*(3), 1–35. https://doi.org/10.1002/rev3.3292

National Centre for Education Statistics. (1996). *Technical issues in large-scale performance assessment*. U.S. Department of Education, Office of Educational Research and Improvement. https://nces.ed.gov/pubs/96802.pdf

National Council of Education Research and Training (NCERT). (2022). Ancient education system in India. In *It so happened: Supplementary reader in English for Class VIII* (Chapter 11, pp. 89–98). https://ncert.nic.in/textbook/pdf/heih111.pdf

Newman, D. B., & Smith, K. (2016). Factors that influence the selection of assessment methods: Recommendations for more purposeful use of online tools. *Journal of Online Learning and Teaching, 12*(1), 1–12. https://jolt.merlot.org/Vol12no1/Newman_0316.pdf.

NSW Education Standards Authority. (n.d.). *Assessment for, as and of learning*. https://educationstandards.nsw.edu.au/wps/portal/nesa/k-10/understanding-the-curriculum/assessment/approaches

Office for Students (2022, July). *Maintaining the credibility of degrees*. Insight brief 14. https://www.officeforstudents.org.uk/media/a712215d-94e8-4e98-8614-baee4d5aa3dc/insight-brief-14.pdf

Orr, G., & Conaway, C. (2016). Time for a change: Four essential strategies for transforming assessment in higher education. *Change: The Magazine of Higher Learning, 48*(5), 42–49. https://doi.org/10.1080/00091383.2016.1227675

Perkins, M., Gezgin, U. B., & Roe, J. (2020). Reducing plagiarism through academic misconduct education. *International Journal for Educational Integrity, 16*(3), 1–15. https://doi.org/10.1007/s40979-020-00052-8

Playfoot, D., Wilkinson, L. L., & Mead, J. (2022). Is continuous assessment inclusive? An analysis of factors influencing student grades. *Assessment & Evaluation in Higher Education, 48*(7), 938–950. https://doi.org/10.1080/02602938.2022.2150834

Popkova, E. (2018). Continuous cumulative assessment in higher education: Coming to grips with test-enhanced learning. In S. Hidri (Ed.), *Revisiting the assessment of second language abilities: From theory to practice* (pp. 309–324). Springer. https://doi.org/10.1007/978-3-319-62884-4_16

Quality Assurance Agency for Higher Education (QAA). (2024, February 19). *Navigating the complexities of the artificial intelligence era in higher education*. https://www.qaa.ac.uk/docs/qaa/news/quality-compass-navigating-the-complexities-of-the-artificial-intelligence-era-in-higher-education.pdf?sfvrsn=8179b281_11

Razı, S. (2023). Emergency remote teaching adaptation of the anonymous multi-mediated writing model. *System, 113*, 102981. https://doi.org/10.1016/j.system.2023.102981

Richardson, K. (2022). What IQ tests test. *Theory & Psychology, 12*(3), 283–314. https://doi.org/10.1177/0959354302012003012

Richardson, K., & Norgate, S. H. (2015). Does IQ really predict job performance? *Developmental Review, 35*, 51–67. https://doi.org/10.1080/10888691.2014.983635

Rowe, N. C. (n.d.). Cheating in online student assessment: Beyond plagiarism. *Online Journal of Distance Learning Administration, 7*(2) Accessed on 27th October 2024 https://www.westga.edu/~distance/ojdla/summer72/rowe72.html

Scott, M., & Unsworth, J. (2018). Matching final assessment to employability: Developing a digital viva as an end of programme assessment. *Higher Education Pedagogies, 3*(1), 373–384. https://doi.org/10.1080/23752696.2018.1510294

Sefcik, L., Striepe, M., & Yorke, J. (2020). Mapping the landscape of academic integrity education programs: What approaches are effective? *Assessment & Evaluation in Higher Education, 45*(1), 30–43. https://doi.org/10.1080/02602938.2019.1604942

Sivasubramaniam, S. D. (2013). Sustainability in student engagement—Assessing the impact of formative 'feed-forward' on reflective learning. In *Proceedings of the HEA STEM learning and teaching conference.* Accessed on 27th October 2024 https://s3.eu-west-2.amazonaws.com/assets.creode.advancehe-document-manager/documents/hea/private/bio-074_0_1568037083.pdf

Sivasubramaniam, S. D. (2014). "Perfecting the plagiarist": Feedback-based continuous assessment strategy to enhance good academic writing. In *Proceedings of the 6th International Integrity & Plagiarism Conference: Promoting authentic assessment (pp. 16–18).* Newcastle.

Shanteau, W. J., Thomas, R. P., & Pounds, J. C. (2002). Performance-based assessment of expertise: How to decide if someone is an expert or not. *European Journal of Operational Research, 136*(2), 253–263. https://doi.org/10.1016/S0377-2217(01)00113-8

Stern, E. (2008). The impact of formative assessment on student motivation, achievement, and academic adjustment. *Review of Educational Research, 78*(1), 153–189. https://doi.org/10.3102/0034654307313795

Stone, E., & Davey, T. (2011, August). Computer-adaptive testing for students with disabilities: A review of the literature. *Educational Testing Service Research Report ETS RR–11-32.* https://files.eric.ed.gov/fulltext/ED523689.pdf

Technavio. (2024). *Anti-plagiarism software for education sector market analysis North America, Europe, APAC, South America, Middle East and Africa–US, China, Germany, Canada, Japan, India–Size and Forecast 2024-2028.* https://www.technavio.com/report/anti-plagiarism-software-market-for-education-sector-industry-analysis

Twist, L. (2021). Changing times, changing assessments: International perspectives. *Educational Research, 63*(1), 1–8. https://doi.org/10.1080/00131881.2021.1876590

Weatherton, M., & Schussler, E. E. (2021). Success for all? A call to re-examine how student success is defined in higher education. *CBE—Life Sciences Education, 20*(1), es3. https://doi.org/10.1187/cbe.20-09-0223

Wiggins, G. P. (1998). *Educative assessment: Designing assessments to inform and improve student performance.* Jossey-Bass.

Wrigley, W. J., & Smith, A. C. (2021). Supporting academic integrity in the age of ubiquitous technology: Lessons learned from a collaboration between IT and academic staff. *Journal of Higher Education Policy and Management, 43*(3), 303–315. https://doi.org/10.1080/1360080X.2020.1863930

Chapter 5
Developing Resources for Supporting Ethical Publishing and Dissemination Activities

Irene Glendinning, Shiva Das Sivasubramaniam, Muaawia Ahmed Hamza, Siavash Miri, Sonja Bjelobaba, and Ana Cristina Veríssimo

Abstract The European Network for Academic Integrity's (ENAI) Ethical Publishing and Dissemination working group has developed guidance targeting early career researchers (ECRs). Materials developed so far cover the responsibilities of different stakeholders in managing academic publishing and a checklist of characteristics that help to distinguish between reputable and disreputable journals and publishers.

Although many types of guidance exist already in different forms, we found no guidance covering all the different publishing roles that is synchronised to provide a comprehensive complementary set. Therefore, we set out to fill that gap.

Despite many existing checklists, the number of disreputable publishers and journals continues to increase, reflecting strong global demand for these services.

I. Glendinning (✉)
Independent Researcher, Leicestershire, UK
e-mail: ireneglendinning@googlemail.com

S. D. Sivasubramaniam
University of Roehampton, London, UK
e-mail: s.sivasubramaniam@roehampton.ac.uk

M. A. Hamza
King Fahad Medical City, Riyadh, Saudi Arabia
e-mail: mahahmed@kfmc.med.sa

S. Miri
Brieflands, Heerlen, Netherlands
e-mail: ceo@brieflands.com

S. Bjelobaba
Uppsala University, Uppsala, Sweden
e-mail: sonja.bjelobaba@uu.se

A. C. Veríssimo
University of Porto, Porto, Portugal
e-mail: averissimo@med.up.pt

M. Davis, C. Baptista (eds.), *Ethics and Integrity in Education (Practice)*, Ethics and Integrity in Educational Contexts 10,
https://doi.org/10.1007/978-3-031-95686-7_5

Other threats come from hijacked journals, paper mills and authorship for sale. Raising awareness in ECRs will improve their understanding about the threats they face and their rights, so they can exercise their responsibilities as an author and make informed decisions on where and how to publish.

In July 2023, the draft guidance notes and checklist were presented at a conference workshop. Participants were asked for feedback to help to validate and improve these resources. This paper will describe the processes followed by working group members to develop and validate these resources, which are available on the working group's website.

Keywords Ethical publishing · Academic publishing; predatory publishing checklist; research integrity · Early career researchers · Peer review guidance

Introduction

The global academic publishing industry is facing unprecedented challenges relating to the variable quality and integrity of services and publications from some publishers and journals. In fact, a recent article by Van Noorden in Nature News demonstrated that in 2023 alone there were more than 10,000 retracted papers (Van Noorden, 2023). This increase in retractions is a complex phenomenon influenced by various factors, including both positive developments like improved detection methods and negative pressures like publication bias and research misconduct (Retraction Watch, n.d.-a, n.d.-b, n.d.-c). The increasing number of retractions of published papers provides some indication of the scale of the problem the industry is facing (Retraction Watch, n.d.-c), but these figures do not take into account the outputs from poor quality, disreputable, predatory publishers and journals, which are published irrespective of quality, with minimum scrutiny and never retracted.

In a similar manner, purveyors of poor-quality services often organise academic conferences, sometimes linked to a guaranteed publication (Eaton, 2020). The conferences are sometimes completely fake, whereby a programme and attendance list are prepared as evidence of the event, no research is presented, but a "conference" publication is generated (for example, see InterConf, n.d.). When there is a physical conference, the event is designed to make money for the organisers, not to provide a forum for an interested audience of researchers to present their own research results and hear about and discuss the research of others in a focused field (Godskesen et al., 2022). Therefore, it is essential for authors and readers to remain vigilant and critically evaluate publishers and journals before engaging with them.

The authors are all members of a working group called Ethical Publishing and Dissemination (EPAD), which is part of the European Network for Academic Integrity (ENAI). We are interested in promoting good practice for all researchers and academics, with a special focus on providing guidance and training for ECRs (EPAD, n.d.). One of the activities of our working group is to develop useful materials to share through the ENAI website.

This chapter is about how the EPAD working group members set about developing useful materials that provide guidance on a range of topics relating to academic publishing. The materials developed so far are guidance notes for peer reviewers and authors and a checklist for helping to distinguish between good quality and disreputable publishers and journals.

Before describing our modus operandi and methodology for developing the materials, we will consider what influences led to the formation of EPAD, and the decision to develop these materials.

Literature Review

Most reputable publishers and journals provide guidance for peer reviewers and authors (e.g.: Taylor & Francis, n.d.; Clarivate, n.d.). However, these are usually specific to that publisher or journal's requirements. ENAI identified the need to acquire or develop guidance for peer reviewers and participants at their annual conferences and for publications relating to the conferences and research. In addition, very little guidance was found that was suitable to advise PhD students and early career researchers (ECRs) on the peer review process. Therefore, EPAD members developed generic guidance notes to match these needs that could be adapted for various purposes, drawing ideas from a range of freely available resources (including: ALLEA, 2023; COPE, n.d.; Nature Editorial, 2022; Taylor & Francis, n.d.).

The rise of open access journals marked a significant shift in the academic publishing landscape. On one hand, they made scholarly research freely available to anyone with an internet connection, removing barriers to access imposed by traditional subscription-based models. However, on the other hand, they resulted in issues regarding quality control (hindering robust peer review), transparency (in editorial processes, and potential conflicts of interest), and financial challenges for authors. Author contributions to article processing charges (APCs) cover the costs of publishing, which can be a barrier for researchers, particularly those from institutions with limited funding sustainability (Björk & Solomon, 2012). Therefore, this leads to inequalities in publishing opportunities, disadvantaging researchers from countries and institutions where less funding is available. Most importantly, the concept of open access has resulted in the proliferation of predatory journals (Beall, 2017; Besir, 2018; COPE, 2019, 2020; Eaton, 2020; Krawczyk & Kulczycki, 2021).

Factors currently affecting the publishing industry are well documented already, including a range of fraudulent practices by publishers, editors, peer reviewers and authors (Banks, 2022; Fenske, 2021). Fraudulent practices by publishers can take various forms, often aimed at misleading authors, readers, or advertisers for financial gain (Gillis, 2018; Glendinning et al., 2019; Grove, 2021). These practices undermine the credibility of academic publishing, eroding trust in scientific research and scholarly communication. Examples of publisher (or fake publisher) related fraudulent practices were well documented by several authors (Leonard et al., 2021; Elmore & Weston, 2020; COPE, 2019; Beall, 2017). These include (but are not confined to) (a) creating journals with little to no peer review process, charging authors exorbitant fees for publication, and falsely claiming academic legitimacy; (b) indirectly engaging (or turning a blind eye) in malpractices such as plagiarism, salami slicing (i.e. publication of two or more articles derived from a single study—Abraham, 2000) to boost their outputs or publishing the same content across multiple journals without disclosure, aiming to inflate publication numbers or manipulate metrics; (c) exploiting the open-access model without any credible editorial team and without any transparency; (d) manipulating or misrepresenting metrics like impact factors, citation counts, or journal rankings to attract authors or advertisers; (e) peer review fraud: fabricating or using fake reviewers to provide positive feedback on submitted manuscripts (bypassing objective peer review); (f) biased editorial practices favouring certain authors subject to financial gains and (g) misrepresenting the reach, readership demographics, or effectiveness of advertising opportunities within publications; (h) charging an author for a citation service, which pays other authors to cite the paper to boost the citation count (Goldacre, 2015; Marcus, 2021).

The drivers behind the urgency for academics to get published and to present their research at conferences are well understood (Abalkina, 2021; COPE, 2019, 2020; Eaton, 2020; Glendinning et al., 2019; Glendinning & Eaton, 2023). Although these drivers are often summarised as "publish or perish", this expression covers a wide range of misplaced initiatives and side-effects from policies designed to influence institutional or national prestige (Glendinning & Eaton, 2023). Factors driving unethical publication practices include (a) paying bonuses to researchers and academics for publishing in high ranking (or sometime any) journals (Prest, 2017); (b) setting a minimum annual publication count for academics in order to stay in post; (c) requiring PhD students to publish one or more peer reviewed journal papers as a condition for completion and graduation; (d) basing academic promotions and appointments on publication counts; (e) incentives designed to increase national and institutional reputation and rankings (Wouters et al., 2019); (f) the time it takes to get accepted and published in a reputable journal; (h) the high APCs levied by many high ranking journals; (i) ignorance about the fraudulent nature of some journals (Eaton, 2020; Glendinning et al., 2019; Glendinning & Eaton, 2023). In addition, fraudsters are taking advantage of the opportunity to make money out of this complex set of circumstances.

Early career researchers (ECRs) are particularly vulnerable to scams, dodgy practices and fraudulent claims by disreputable publishers, unless they receive training and guidance about how to identify and respond to any threats that may arise. However, many other academics knowingly submit to disreputable journals in order to survive in the challenging environment of academia (Abalkina, 2021; Glendinning et al., 2019).

Many other guidance notes, checklists, blacklists and whitelists exist. The most well-known of the blacklists was created and maintained by Jeffrey Beall, until he retired from his role as librarian at University of Denver in 2017 (Beall, 2017). The list still exists, both in the final form in which it was left by Jeffrey Beall and has also continued to be maintained by an unknown party (Beall's List, n.d.). The other widely used blacklist and whitelist is Cabells (n.d.), which is a subscription service. Cabells also share the criteria they use to evaluate journals and publishers.

A range of other sources are freely available that provide useful information, tips and lists of criteria that characterise poor quality, disreputable journals and publishers (including: Abbott, 2017; ALLEA, 2023; Bridge Project, n.d.; Clarivate, n.d.; COPE, 2020; DOAJ, n.d.; Eaton, 2020; Eaton & Carmichael, 2023; Elmore & Weston, 2020; Glendinning, 2024; Hunziker, 2017; Leonard et al., 2021; McCook, 2016; Retraction Watch, n.d.-a, n.d.-b, n.d.-c; Think Check Submit, n.d.; Vilhelmsson, 2022).

Methodology

Generic guidance for authors and peer reviewers was developed by EPAD several years ago by referring to and learning from content in guidance documents from various sources (for example, COPE, 2019; Nature, 2024; Wiley Author Services, n.d.). The guidance notes were then applied and further developed from direct experience in the context of several of the annual conferences organised by ENAI and for preparing publications associated with ENAI events and research.

Working group members identified a wide range of sources that identify criteria relating to predatory and disreputable publishing. Most of the sources collected are included in the literature review of this chapter. One of the authors then used ChatGPT to summarise and classify the criteria from the selected sources. Members of the working group then commented on the document, corrected mistakes, improved formatting, grammar and wording and suggested ways to more logically sequence and group the content.

The resulting draft was presented to a workshop at the ECEIA conference in Derby in July 2023. A leaflet was designed, printed and circulated during the conference by one of the authors to publicise the workshop and to invite feedback.

About 15 workshop participants were asked to work in small groups, first reading through at least one section of the checklist, then to discuss and provide (anonymous) feedback on:

- What is helpful, new, unexpected;
- Anything missing that should be added;
- Anything unclear;
- Things you disagree with;
- Other ideas to improve the checklist.

Participants were also invited to provide (anonymous) feedback on the reviewer and author guidance notes after the conference.

QR codes were provided for participants to access the checklist, the peer reviewer guidance and the author guidance. The QR codes and URLs were also made available on the PowerPoint presentation. In addition, coloured copies of the draft checklist were printed for use by participants, highlighting positive (green), negative (red) and neutral (orange) criteria about journals and publishers on the checklist. A Google form version of the checklist was also made available to participants.

Following the conference, working group members responded to the feedback provided by editing the checklist. Many of the comments concerned duplication of content and a few errors were found in categorisation of items. Two of the authors then worked on ways to remove the colour coding, by grouping positive and negative factors within each section of the guidance. The final stage was to add a scoring system to the checklist.

After the checklist was made available on-line early in 2024, we invited feedback. As a result, a researcher in this field contacted the EPAD leader, questioning the accuracy and relevance of some of the items included on our checklist. The team responded to the critique with some minor changes to wording. In addition, the researcher suggested that we reviewed their publications, the most recent of which included a much shorter checklist than ours (Kratochvil et al., 2019; Kratochvil et al., 2020). Our responses to this useful feedback are detailed in the following discussion section.

Discussion

All the materials generated by EPAD are available for reuse and adaptation under a Creative Commons CC BY NC License (ENAI, n.d.).

Guidance Notes Relating to Peer Review

There are four sets of guidance notes available via the ENAI web site at present relating to peer reviewing for conferences. There is an overview of different roles and responsibilities relating to peer review (EPAD, n.d.). Sets of generic guidance notes for reviewers and authors are both available for downloading and reuse (EPAD, n.d.). One other document with guidance for conference session chairs that was developed for a recent conference is also available from the same web page (EPAD, n.d.). The EPAD team will continue to refine and improve these documents and create further related documents, including guidance for reviewing full conference papers. The longer-term plan is for EPAD members to develop videos to guide inexperienced reviewers and authors on the peer review process.

The Purpose of the Checklist

The value of checklists and guidance notes for explaining publishing practices and for evaluating the quality of journals and publishers, depends on the accuracy and relevance of the items listed, how they are categorised, organised and how easy they are to use. Many websites and resources for listing, ranking and/or evaluating journals and publishers (including DOAJ, n.d.; Scimago, n.d.; Clarivate, n.d.) have been criticised for giving respectability to journals and publishers that some researchers believe to be "potentially predatory" (Bik, 2024; Shamseer et al., 2017; Vinny et al., 2016). Even more worryingly, in recent years the range of checks and quality processes in many highly respected and high-ranking journals and publishing giants have failed to detect and prevent the publication of poor quality and fabricated papers (Bohannon, 2013; Van Noorden, 2023; Retraction Watch, n.d.-a, n.d.-b, n.d.-c).

As explained above, after we made the checklist publicly available via the ENAI website, we received some valuable feedback from a member of another team researching in this field (Kratochvil et al., 2019; Kratochvil et al., 2020), which led to some minor wording changes in our checklist. The first paper by Kratochvil et al. was an empirical study of 259 open access biomedical journals indexed in the Web of Science and Journal Citation Reports (WoS/JCR), testing for their compliance with 16 whitelist criteria signalling "transparency and best practice in scholarly publishing" (Kratochvil et al., 2019, p.338). Only four journals out of 259 in the study met all the criteria (Kratochvil et al., 2019, p. 342). As all other checklists contain points that also apply to many journals, otherwise considered to be reputable and listed as such in WoS/JCR, Scopus and MEDLINE, one conclusion from their 2019 paper was that these points are unsuitable for use in any whitelist/checklist.

The second paper by the same team started with a list of 19 criteria and after further evaluation of their analysis, reduced this to a list of 10 criteria after finding that 74 journals "may have been incorrectly assessed as untrustworthy using the longer list" (Kratochvil et al., 2020, p. 308). Based on their own research, the claim was made by the lead author that our checklist is unreliable.

After some consideration of this research and their assertions, we decided not to make further changes to our checklist at this point, but we remain open to discussion and ideas and will continue to communicate with this team of researchers. The reasoning behind this decision is that our checklist was not designed to be a whitelist that categorises journals and publishers as reputable or a blacklist that identifies disreputable journals. It is intended as a guide for authors, particularly to support ECR academics new to publishing, on factors to consider when selecting a journal for their own publications. We still believe that all the criteria we include in our checklist have a part to play in supporting its original function. Our scoring method allows for journals that may have weaknesses based on some criteria that are not overall untrustworthy.

Using the Checklist

The checklist is currently available for download from the ENAI website (EPAD, 2024). There are plans to create an interactive ENAI web page that will allow people to enter details of journals and publishers, together with their characteristics, that will automatically calculate the score and make recommendations. The checklist has eight Categories, each containing between three and eleven factors. Table 5.1 below is the scoring grid from the Checklist, showing the eight Categories.

As can be seen from Table 5.1, the checklist currently contains 61 factors to check, each with a score, depending on whether the factor is a positive (1) or negative (0) indicator of publishing ethics and quality. A score of 30 or above would normally indicate a reputable journal, but a below-threshold score on any of the eight categories will override the overall decision.

The next stage of the development is to further test the accuracy and completeness of all the resources and the efficacy of the checklist and scoring process. Feedback has already been received, duly considered and acted upon in the short time since the checklist was made available on the ENAI web site. Further development of the checklist will consider whether different weightings or negative scoring should be assigned to the more critical factors, such as promising a five-day turnaround from submission to publication, including rigorous peer review and guaranteed publication. We are particularly interested in further exploring the ten critical criteria proposed by Kratochvil and colleagues, to determine whether these should be classified or weighted differently in our checklist.

Table 5.1 The scoring grid for the predatory publishing checklist

	Category	Max score	Sub-total	Threshold for OK	Notes on weaknesses
A	Journal title and authenticity	6		> = 3	
B	Financial aspects	10		> = 6	
C	Membership and associations	3		> = 2	
D	Quality and scope	8		> = 4	
E	Indexing and archiving	9		> = 5	
F	Editorial and publishing practices	10		> = 6	
G	Communication and invitations	11		> = 6	
H	Copyright and legal aspects	4		> = 2	
Total score and decision		61		> = 30	

Conclusions

Since the launch of the first scientific journal in the seventeenth century, scientists around the globe have opted to publish their research findings in scientific journals. Earlier forms of scientific review were largely based on communications between trusted experts in the same field, before the modern concept of peer review was established in the 1970s (MIT Communication Lab, n.d.). Until the 1990s, peer reviewed scholarly journals had been viewed as the most trusted source of information for sharing knowledge and research results in different disciplines. However, after the open access publication movement started to develop, the scientific publication landscape was transformed. In 2012, Professor Peter Suber of Harvard University defined and advocated OA publications by presenting the benefits (Suber, 2012). Unfortunately, amidst the celebrations by the scientific community of the advent of OA, different forms of fake journals emerged that exploited this new era of scientific publishing and all stakeholders, in many different ways, as explained in this chapter.

In addition to the rise in disreputable publishers and editors, this chapter also discussed different ways that the integrity and reputation of the scholarly publishing industry as a whole is being undermined through corruption and fraud by researchers, authors, peer reviewers and paper mills (Abalkina, 2021; Abraham, 2000; Meers, 2018; Nolfi et al., 2015; Oransky, 2019). The emergence of fake journals raised concerns of academics, researchers, and the general public, over the believability and reliability of claims in scientific publications, with clear justification. It did not take long before the scientific community was mobilised to act against these cancerous phenomena. Although the factors that encourage and drive authors to publish in fake journals are well documented (Prest, 2017; Vilhelmsson, 2022; Wouters et al., 2019, and many more publications), very little has been done so far to remove misguided incentives and otherwise reduce the demand for publishing in disreputable and fake academic journals.

The authors of this chapter and their colleagues in ENAI's EPAD working group have dedicated their time to educate and raise awareness in scholars through events, conferences and in different institutions in almost all continents of the world. EPAD members have specifically focused on educating early career researchers, who are the most vulnerable to falling in the trap of publishing in fake journals (as discussed in: Glendinning et al., 2022; Bridge Project, n.d.). This guidance and the checklist, like other publishing guidance developed in different disciplines, are resources and tools designed to help educate, evaluate and protect ECRs, and other researchers, from victimisation by fake publishers. Importantly, the authors are aware that this type of guidance is useful to anyone involved or interested in academic publishing, not just inexperienced researchers.

The EPAD group took the initiative to prepare and present guidance notes and the checklist at the annual ENAI conference in July 2023 in Derby, UK. In the short time since the checklist was made publicly available, further research in this field has already emerged that we can learn from in order to improve the utility and functionality. All suggestions, ideas and comments to date have been taken in consideration to improve the checklist.

The resources developed by EPAD are available for general access via the ENAI website. A programme of active dissemination and promotion is achieved through regular workshops on ethical publishing at summer schools and other events as part of on-going activities by EPAD members and through ENAI and related projects.

The benefit of having a checklist is that it can always be updated and improved, which EPAD plans to continue to do. The authors of this chapter believe that this checklist will not only help ECRs to avoid publishing in fake journals, but also encourage more research in this field. In particular, it would be valuable to conduct several research studies across different cultures to evaluate how useful the checklist is. The authors and EPAD members very much welcome further constructive feedback on any of the resources that will help to improve them, for the benefit of all.

References

Abalkina, A. (2021). *Publication and collaboration anomalies in academic papers originating from a paper mill: Evidence from a Russia-based paper mill*. ArXiv, 2112.13322v2. https://doi.org/10.48550/arxiv.2112.13322

Abbott, J. H. (2017). How to choose where to publish your work. *Journal of Orthopaedic & Sports Physical Therapy, 47*(1), 6–10. https://doi.org/10.2519/jospt.2017.0102

Abraham, P. (2000). Duplicate and salami publications. *Journal of Postgraduate Medicine, 46*(2), 67–69. https://pubmed.ncbi.nlm.nih.gov/11013467/

ALLEA, All European Academies. (2023, June). *The European code of conduct for research integrity*. https://allea.org/code-of-conduct/

Banks, M. (2022, July 27). University of Fukui professor called out for fake peer review, loses "love hormone" paper. *Retraction Watch*. https://retractionwatch.com/2022/07/27/university-of-fukui-professor-called-out-for-fake-peer-review-loses-love-hormone-paper/

Beall, J. (2017). What I learned from predatory publishers. *Biochemia Medica, 27*(2), 273–278. https://doi.org/10.11613/BM.2017.029

Beall's List. (n.d.). *Beall's list of potentially predatory journals and publishers*. https://beallslist.net/

Besir, D. S. (2018). Predatory journals: Who publishes in them and why? *Journal of Infometrics, 12*(4), 1296–1311. https://doi.org/10.1016/j.joi.2018.10.008

Bik, E. (2024, February 15). *The rat with the big balls and the enormous penis—How Frontiers published a paper with botched AI-generated images. Science Integrity Digest.*. https://scienceintegritydigest.com/2024/02/15/the-rat-with-the-big-balls-and-enormous-penis-how-frontiers-published-a-paper-with-botched-ai-generated-images/

Bohannon, J. (2013). Who's afraid of peer review? *Science, 342*(6154), 60–65. https://doi.org/10.1126/science.342.6154.60

Bridge Project. (n.d.). *Building integrity in higher education, business and society*. European Network for Academic Integrity. https://www.academicintegrity.eu/wp/bridge/

Björk, B. C., & Solomon, D. (2012). Open access versus subscription journals: A comparison of scientific impact. *BMC Medicine, 10*, 73. https://doi.org/10.1186/1741-7015-10-73

Cabells. (n.d.). *Cabells scholarly analytics*. https://www2.cabells.com/about-predatory

Clarivate. (n.d.). *Web of science*. https://clarivate.com/webofsciencegroup/solutions/web-of-science/

COPE, Committee on Publication Ethics. (n.d.). *Publication guidelines*. https://publicationethics.org/guidance/Guidelines

COPE, Committee on Publication Ethics. (2019, February 23). *Systematic manipulation of the publication process*. https://doi.org/10.24318/cope.2019.2.23.

COPE, Committee on Publication Ethics. (2020). *Discussion document: Predatory Publishing*. https://publicationethics.org/guidance/Guidelines?t=COPE+%282020%29.+Discussion+document%3A+Predatory+Publishing.+&sort=score

DOAJ. (n.d.). *Directory of open access journals*. https://doaj.org/

Eaton, S. E. (2020). Avoiding predatory journals and questionable conferences: A resource guide. *University of Calgary.*. https://prism.ucalgary.ca/server/api/core/bitstreams/3c1b1ccb-e82b-49c3-96e5-7d4f3fcb9352/content

Eaton, S. E., & Carmichael, J. J. (2023). Fake degrees and credential fraud, contract cheating, and paper mills: Overview and historical perspectives. In S. E. Eaton, J. J. Carmichael, & H. Pethrick (Eds.), *Fake degrees and fraudulent credentials in higher education. Ethics and integrity in educational contexts* (Vol. 5, pp. 1–22). Springer. https://doi.org/10.1007/978-3-031-21796-8_1

Elmore, S. A., & Weston, E. H. (2020). Predatory journals: What they are and how to avoid them. *Toxicologic Pathology, 48*(4), 607–610. https://doi.org/10.1177/0192623320920209

ENAI, European Network for Academic Integrity. (n.d.). *Homepage*. https://www.academicintegrity.eu/wp/

EPAD, Ethical publishing and Dissemination. (n.d.). *ENAI working groups*. https://www.academicintegrity.eu/wp/ethical-publishing-and-dissemination-working-group/

EPAD. (2024, April 15). A guide to identifying predatory journals and publishers–checklist. *ENAI resources*. https://www.academicintegrity.eu/materials/277

Fenske, J. (2021, December 30). Unreliable science in media as peer-review breaks down. *The Click*. https://theclick.news/unreliable-science-in-media/

Gillis, A. (2018. March 5). *Poor-quality, Predatory conferences prey on academics*. University Affairs/Affaires Universitaires (UA/AU). https://www.universityaffairs.ca/news/news-article/poor-quality-predatory-conferences-prey-academics/

Glendinning, I., Ribeiro, L., Sivasubramaniam, S., Ahmed, M. A., Gholami, G., & Marar, S. D. (2022). *Educating early career researchers on publishing options* (pp. 299–309. Retrieved from). Paper presented at IRAFPA Colloquium. https://doi.org/10.56240/cmb9927

Glendinning, I., & Eaton, S.E. (2023). Understanding and addressing drivers of corruption in academic publishing. In Eaton, S.E, Handbook of academic integrity. (Vol II, Part VIII, chapter 83, pp. 1403–1422). Springer. doi:https://doi.org/10.1007/978-981-287-079-7_146-1.

Glendinning, I., Orim, S., & King, A. (2019). *Policies and actions of accreditation and quality assurance bodies to counter corruption in higher education*. CHEA/CIQG 2019. Executive summary, full report: https://www.chea.org/corruption-higher-education

Glendinning, I. (2024). Addressing systemic forms of corruption affecting educational standards and quality. In E. Denisova-Schmidt, P. Altbach, & H. de Wit (Eds.), *Handbook on corruption in higher education*. Edward Elgar Publishing. in press.

Goldacre, B. (2015, August 14). So this company is paying authors for citations in academic journals. *Bad Science*. https://www.badscience.net/2015/08/so-this-company-cyagen-is-paying-authors-for-citations-in-academic-papers/

Godskesen, T., Eriksson, S., Oermann, M. H., & Gabrielsson, S. (2022). Predatory conferences: A systematic scoping review. *BMJ Open, 12*(11), e062425. https://doi.org/10.1136/bmjopen-2022-062425

Grove, J. (2021, August 30). Hijacked Journals 'siphon millions of dollars' from research. *Times Higher Education*. https://www.timeshighereducation.com/news/hijacked-journals-siphon-millions-dollars-research

Hunziker, R. (2017). Avoiding predatory publishers in the post-Beall world: Tips for writers and editors. *AMWA Journal, 32*(3), 113–115. https://cdn.ymaws.com/www.amwa.org/resource/resmgr/journal/Spotlight/2017v32n3_AvoidingPredatoryP.pdf

InterConf. (n.d.). *Interconf scientific publishing center*. https://interconf.center/en_interconf-switzerland.html?spush=aXJlbmVnQGNvdmVudHJ5LmFjLnVr

Kratochvil, J., Plch, L., & Koriťáková, E. (2019). *Compliance with ethical rules for scientific publishing in biomedical open access journals indexed in journal citation reports*. Vnitř Lék 2019; 65(5): pp. 338–347. University Campus Library, Masaryk University Brno. http://eprints.rclis.org/38394/

Kratochvil, J., Plch, L., Sebera, M., & Koriťáková, E. (2020). Evaluation of untrustworthy journals: Transition from formal criteria to a complex view. *Learned Publishing, 33*(3), 308–322. https://doi.org/10.1002/leap.1299

Krawczyk, F., & Kulczycki, E. (2021). How is open access accused of being predatory? The impact of Beall's lists of predatory journals on academic publishing. *The Journal of Academic Librarianship, 47*(2), 102271. https://doi.org/10.1016/j.acalib.2020.102271

Leonard, M., Stapleton, S., Collins, P., Selfe, T. K., & Cataldo, T. (2021). Ten simple rules for avoiding predatory publishing scams. *PLoS Computer Biology, 17*(9), e1009377. https://doi.org/10.1371/journal.pcbi.1009377

Van Noorden, R. (2023, December 12). More than 10,000 research papers were retracted in 2023—A new record. *Nature News*. https://www.nature.com/articles/d41586-023-03974-8#:~:text=Rising%20rates,2022%2C%20it%20exceeded%200.2%25.

Marcus, A. (2021, August 13). Publisher offers cash for citations. *Retraction Watch*. https://retractionwatch.com/2021/08/31/publisher-offers-cash-for-citations/.

McCook, A. (2016, October 24). 7 signs a scientific paper's authorship was bought. *Retraction Watch*. https://retractionwatch.com/2016/10/24/seven-signs-a-paper-was-for-sale/

Meers, J. (2018, January 20). *Dissernet: Exposing fraud and Plagiarism in Russia's Academia*. Investigative journalism education consortium. https://ijec.org/2018/01/20/dissernet-exposing-fraud-and-plagiarism-in-russias-academia/

MIT Communication Lab. (n.d.). *Peer review—a historical perspective*. https://mitcommlab.mit.edu/broad/commkit/peer-review-a-historical-perspective/

Nature Editorial. (2022). *Nature is trialling transparent peer review–the early results are encouraging*. https://www.nature.com/articles/d41586-022-00493-w

Nature. (2024). *For authors*. https://www.nature.com/nature/for-authors

Nolfi, D. A., Lockhart, J. S., & Myers, C. R. (2015). Predatory publishing: What you don't know can hurt you. *Nurse Educator, 40*(5), 217–219. https://doi.org/10.1097/NNE.0000000000000179

Oransky, I. (2019, December 27). *Disbelief: Researchers, watch out for this new scam involving journal special issues*. Retraction Watch. https://retractionwatch.com/2019/12/27/disbelief-researchers-watch-out-for-this-new-scam-involving-journal-special-issues/

Prest, K. (2017, July 24). *Analysis of Chinese universities' financial incentives for academic publications*. British Council. https://opportunities-insight.britishcouncil.org//insights-blog/analysis-of-chinese-universities%E2%80%99-financial-incentives-academic-publications.

Retraction Watch. (n.d.-a). *Highly cited engineer offers guaranteed publication, citations in return for coauthorship*. https://retractionwatch.com/

Retraction Watch. (n.d.-b). *The retraction watch leaderboard*. https://retractionwatch.com/the-retraction-watch-leaderboard/

Retraction Watch. (n.d.-c). *Retraction watch database user guide*. https://retractionwatch.com/retraction-watch-database-user-guide/

Scimago. (n.d.). *SJR–Scimago journal and country rank*. https://www.scimagojr.com/

Shamseer, L., Moher, D., Maduekwe, O., Turner, L., Barbour, V., Burch, R., . . . Shea, B. J. (2017). Potential predatory and legitimate biomedical journals: can you tell the difference? A cross-sectional comparison. *BMC Medicine, 15*(1), 28

Suber, P. (2012, July 20). *Open Access* . MIT Press Essential Knowledge Series.

Taylor & Francis (n.d.). *A guide to becoming a peer reviewer*. https://editorresources.taylorandfrancis.com/reviewer-guidelines/

Think Check Submit. (n.d.). *Identify trusted publishers for your research*. https://thinkchecksubmit.org/

Vilhelmsson, A. (2022, April 14). Predatory publishing 2.0: Why it is still a thing and what we can do about it. *PLOS Blogs: ECR Community*. https://ecrcommunity.plos.org/2022/04/14/predatory-publishing-2-0-why-it-is-still-a-thing-and-what-we-can-do-about-it/

Vinny, P. W., Vishnu, V. Y., & Lal, V. (2016). Trends in scientific publishing: Dark clouds loom large. *Journal of the Neurological Sciences, 363*(Supplement C), 119–120. https://doi.org/10.1016/j.jns.2016.02.040

Wiley Author Services. (n.d.). *Review our manuscript preparation guidelines*. https://authorservices.wiley.com/author-resources/Journal-Authors/Prepare/manuscript-preparation-guidelines.html/index.html

Wouters, P., Sugimoto, C. S., Larivière, V., McVeigh, M. E., de Pulverer, B., Rijcke, S., & Waltman, L. (2019). Rethinking impact factors: Bbetter ways to judge a journal. *Nature Comment, 569*, 621–623. https://www.nature.com/articles/d41586-019-01643-3?WT.ec_i

Chapter 6
A Community of Practice Approach to Supporting Academic Integrity in Irish Higher Education

Elva Casey

Abstract The challenges faced by Institutions of Higher Education in identifying, supporting, facilitating, and enhancing good academic practices are growing exponentially. This is due in part to rapid and unprecedented advances in Generative Artificial Intelligence (GenAI) and also to the speedy manner in which these have become part of the parlance around academic integrity. This chapter seeks to address how academic institutions can embed collaborative practices into their planning for and responses to these developing circumstances, in a dynamic and inclusive fashion. The purpose of the chapter is to advocate for a community-based approach to a community issue and to present an example of such an approach within the context of an academic institution of higher education.

The world of academia is grappling with sudden widespread public interest in academic misconduct and the multitude of ways to cheat and 'cheat well'. While institutions move at different paces and with varying degrees of caution to regulate assessments, develop policies and initiate procedures to support best practice, students are faced with increasing temptations to take shortcuts to academic attainment. Those temptations are particularly alluring to students who may be classified as 'vulnerable' (e.g. students entering Higher Education with non-traditional qualifications or from non-traditional routes). Although the potential for abuse, misuse, or avoidance of the benefits of AI is a cause for trepidation across all areas of academia, it is arguably of graver concern in relation to professional qualifications, where shortcuts in assessment responses may lead to the dilution of professional competencies. This chapter recounts the Community of Practice (CoP) based approach to these new challenges developed by an Institution of Higher Education with a specific focus on professional qualifications in teaching and nursing. In doing so, it presents some of the initial evidence of the benefits of this approach including increased student agency specific to academic practices, improved sustained collaborations between faculty, staff and students and the facilitation of an ongoing

E. Casey (✉)
Maynooth University, Maynooth, Ireland
e-mail: elva.casey@mu.ie

M. Davis, C. Baptista (eds.), *Ethics and Integrity in Education (Practice)*, Ethics and Integrity in Educational Contexts 10,
https://doi.org/10.1007/978-3-031-95686-7_6

academic integrity review cycle. The CoP forms part of a cross departmental collaborative approach to creating workable and sustainable practices and policies and as such is replicable across a wide range of institutions. The aim of this chapter is also to demonstrate how to move the focus away from punitive penalties and towards enabling and recognising good academic practice, particularly in relation to students who have experienced challenges to date with their academic journeys.

Keywords Community of Practice · GenAI · Academic integrity

Introduction

The subject of academic integrity and the promotion of cultures of good and sustainable academic practices has moved from the realm of academics and administrators to the public forum in recent years. This is due in part to widespread and often sensationalist media coverage of advances in Generative Artificial Intelligence (GenAI), particularly the launch of ChatGPT in 2022. While the challenges around promoting, supporting, and assuring academic integrity in the face of GenAI are growing, the recognition of academic integrity as an integral aspect of the academic practices of institutions of higher education is also on the rise. Consequently, opportunities for and challenges to academic integrity can be viewed as two sides of the same GenAI coin. GenAI presents an opportunity for policy makers, academics, administrators and the 'traditional' drivers of academic integrity to put academic integrity firmly on the reform agenda.

The European Union categorises the use of AI (artificial intelligence) in education as 'high risk,' addressing the capacity to do harm (European Union, 2024). However, alongside this risk-based approach is an acknowledgement of the potential of AI as a tool for personalised learning, an enhancer of formative assessment, a provider of educational administration and potentially a source of real-time feedback and individualised responses to learners on their learning trajectory. Therefore, educators and learners function within a space where the potential for advancement and the challenges to good academic practice are in a state of constant tension. This chapter proposes harnessing the energy of tension, in other words leaning into the debate, to encourage engagement, sustained conversations and a community-based approach to supporting academic integrity. This approach conforms to International Center for Academic Integrity (ICAI) efforts to frame academic integrity in terms which are both positive and pragmatic (ICAI, 2021).

Academic Integrity involves commitment to the values of honesty, truth, fairness, respect, responsibility, and courage (ICAI, 2021). Packalen and Rowbotham (2022) argue that for policies and procedures in academic integrity to be impactful, a move away from generalised concerns and towards a mutual recognition by faculty, administrators, and students that there is a problem in the status quo, is required. All stakeholders must then, they attest, be in possession of a shared motivation to solve that problem. Academic integrity is more than the avoidance of academic misconduct. It starts with a shared appreciation of the benefits of good academic practices at a societal level as well as at an institutional and individual level. When academic integrity is reduced to an overly simplistic interpretation it is likely to continue to remain within the domain of the 'enforcers' rather than engage the interest of all stakeholders, including students, in the creation of cultures of integrity. Moriarty and Wilson (2022) state that for academic integrity to enhance cultures of integrity and justice, a balance of the pursuit of fairness and justice with humanity, compassion, and empathy is essential. This rationale, that academic integrity belongs to the culture and community of institutions, informs the approach outlined in this chapter. This chapter argues that for the values of academic integrity to become embedded in an institution, they must become part of the community values and therefore be engaged with by a wide cross-section of the institution's community.

Unfortunately, policies and procedures informing and guiding academic integrity within institutions are often perceived as inaccessible and confusing by the very students who need to engage with them (Reedy et al., 2021). Tatum (2022) argues that the use of honour codes in devising policies on academic integrity is only beneficial when interwoven with the institutional culture and further connected to the learning and growth outcomes of students. This leads to the vision of a system that supports all students, particularly those who are vulnerable. Even when students contravene policies or act without integrity, the intent is to continue to help and support the students back onto the pathway to academic success.

This chapter charts the development of a Community of Practice (CoP), within an Institute of Higher Education, delivering programmes of professional qualifications. The CoP is aimed at fostering a mutual, collaborative and sustainable approach to the emerging challenges to academic integrity, in the face of GenAI. The primary purpose of the CoP is to foster a network within the institution to enable culture enhancement, with a specific focus on the need to support and include more vulnerable students. It further highlights some of the outputs to date of the CoP including the development of an Academic Integrity Champions Network, new policies, and principles to guide best practice in academic integrity and the development of the Case Investigator role.

The Institutional Context

Hibernia College is a private Institution of Higher Education based in the Republic of Ireland. It is a supplier of Quality and Qualifications Ireland (QQI) validated programmes of education and provides programmes of professional qualifications in teacher education and nursing. As the largest single source of initial teacher training at primary level in the country, and with a growing portfolio of online and blended programmes, the college has significant experience of designing innovative and authentic assessments and planning for the challenges of less traditional and primarily online forms of assessment. However, it is a widely held perception, supported by some evidence, that online assessment represents a threat to academic integrity and is more susceptible to cheating than traditionally invigilated exams (Mate & Weidenhofer, 2021). The college has therefore, since inception, had a strong focus on assuring robust practices in academic integrity to negate against these potential risks.

As a college with a non-traditional profile and a strong focus on adaptive and transformative learning, Hibernia College is well positioned to consider the challenges and opportunities of advances in technology to the delivery and assessment of programmes of education. With the opportunities for innovation presented by the pandemic resulting in advances such as the piloting of an integrative authentic virtual scenario-based learning environment to support assessment of programme learning outcomes on professional programmes, the college has taken significant steps towards researching and implementing technological advances in education.

However, a cautious and risk-based approach to the use of artificial intelligence will be necessary (EU Artificial Intelligence Act, 2024), to mitigate against adding to the confusion and anxiety already felt by academic faculty, administrative staff, and students in detangling the plethora of information and misinformation around GenAI, its capabilities and its limitations. This project was developed from a recognition of the emerging challenges to academic integrity alongside an acknowledgement of the potential uses of artificial intelligence in providing adaptive learning and individualised content (Kabudi et al., 2021). It was also deemed important from the beginning to avoid paralysis of perfection and becoming caught in an endless 'discuss, pause, repeat' cycle.

Therefore, a guiding initial principle of the project as stated in the Terms of Reference for the Community of Practice was *"to take an iterative approach to collaboratively developing a culture of integrity within Hibernia College, with the potential to address emerging complex ethical and assessment standards challenges."* (Hibernia College, 2024).

The use of the word 'potential' is both significant and purposeful in this context. It is recognised by Hibernia College that what constitutes 'best practice' and 'sound theory' is changing rapidly in this area. Policies that present as fit for purpose at a specific point in time may, indeed will, require review within a short timeframe. Therefore, an approach to policy development which prioritises solutions is doomed to failure. For this reason, the approach outlined here is guided by the aim of keeping collaborative channels of communication open, making small but impactful changes, and accepting that review and self-evaluation are critical to the process.

Cultures of Academic Integrity

A National Culture of Academic Integrity

The national context and strategic approach in Ireland is described by Hackett et al. (2024) as aimed at fostering a culture of academic integrity through a combination of developing key resources, engaging with leaders in higher education, encouraging involvement in the National Academic Integrity Network (NAIN) and supporting a national vision of well-informed, educated, and aware stakeholders. This national context is ably facilitated by NAIN, a peer network of academic and professional staff, student representatives and representative agencies. NAIN has published, through collaborative working groups with representation across NAIN membership, resources to establish a common lexicon of language around academic integrity (QQI, 2021), aid investigate processes (QQI, 2023) and provide guidance on Generative Artificial Intelligence (GenAI) to educators. Therefore, a national context for the establishment of cultures of integrity exists in Ireland. Prior to establishing an institutional community-based approach to academic integrity, it is advisable to consider the national and European context and where synergies can be created between institutional language, actions and policies and these contexts.

An Institutional Culture of Academic Integrity

At an institutional level, a broad definition of institutional culture is a commonly arrived at sense of reality (Vaughan, 1994). Campbell and Hourigan (2008, p. 37) describe negotiating culture values in institutions of higher education as the establishment of "equitable and egalitarian policies and procedures, valuing confrontation, interest groups, mediation and power." Culture is generally accepted to be socially constructed and inclusive of dominant or prevailing beliefs, ideologies, and shared assumptions (Adams, 2014). There is no one universally accepted definition of academic culture, however Tikhonova and Raitskaya (2023) describe a triad of organisational cultures, educational cultures and research cultures forming the fundamental basis for academic cultures. These academic cultures form part of complex ecosystems, based in policies, principles, and established practices. It follows that for policies to be enacted meaningfully and embraced by the organisations they serve; they must be intrinsically and visibly linked to the institutional culture and ideally constructed through a consultative process.

The fundamental values of academic integrity, introduced above, (ICAI, 2021) are core to academic cultures. ICAI (2021) also describes academic integrity cultures as flourishing when members of academic communities live the values. Living values that are espoused by institutions necessitates that people understand and feel both supported and enthused by those values. The values must function as motivational principles to guide attitudes, norms and actions (Davidov et al., 2008). In order for this to happen, responsibility needs to be taken by members of the College community to role model the lived values in action. This can include faculty and staff sharing experiences and case studies that highlight the importance of integrity, thereby inspiring students to adopt these values in their own academic pursuits.

An error that can be made is to assume that academic integrity is an absolute. How people respond to academic challenges can vary significantly in different contexts and cultures. Çelik and Razi (2023) make the significant point that academic integrity should be addressed explicitly by institutions as a pedagogical responsibility. By doing so, the specific challenges of a particular module or course of study can be addressed. Kier and Ives (2022) conclude from their research that a balanced approach to addressing policies, procedures, compliance, commitment, and resources is essential to meet the challenges to academic integrity. The picture that emerges is one whereby no singular static approach will be successful, but rather an agile framework set firmly within the institutional culture is required.

Communities of Practice (CoP)

The term 'Community of Practice' has gained significant popularity in academic settings since first coined by Lave and Wenger (1991). CoPs focus on situated learning through participation in a community. Within an academic integrity context, experimentation in the use of CoPs has focused on facilitating support groups to discuss best practice, nurturing professional growth, and providing safe spaces to discuss challenges (Eaton et al., 2021). This represents a welcome shift away from a focus on punishments and threats of exposure towards more inclusive practices. Draper and Fisher (2020) present a case study of students and staff engaging in partnership to develop a practical toolkit to support directed and independent learning by students. The project has positive impacts in building student confidence, wellbeing, sense of satisfaction and achievement. It is based on the principle that students are best placed to provide solutions collaboratively with faculty. Reedy et al. (2021) recommend avoiding a top-down or didactic approach to policy review and analysis with students as passive recipients of outcomes. A CoP approach to policy design and development therefore can focus on posing and dialoguing questions, rather than imposing barriers. These questions include asking what should the characteristics of the policy be and by what should they be guided? This enables the holistic development and maturation of policies appropriate to cultures and values which are consequently more sustainable and enabled to flourish.

The Challenges for a CoP

While theoretically a sound model, there is a danger with CoPs that they can lose momentum, lack focus and become talking exercises. This can result from a lack of required outputs. Although too much focus on outcomes can impede discussion and collaboration, sustaining interest is challenging when success is difficult to measure.

Identifying the best structure for a CoP for the purposes of promoting and supporting academic integrity is challenging. In this chapter, the 5Ps approach taken by Hibernia College, 'plan, predict, preview, pilot, pivot' is recommended. In the first instance, key senior members of the Department of the Registrar and Academic Faculty met with students to plan a broad session structure and to outline potential content for the first CoP session based solely on participant interests. Next, an effort was made to predict potential key factors at a national and institutional level which will impact on the College over the next 12 months and to integrate those into the workplan. Following on from this, a small CoP was conducted with limited membership to preview the process and work out any logistical and practical issues which might need to be addressed such as timetabling, issuing invites, facilitating online and recording content. Taking the learning from these experiences on board, the CoP was piloted and feedback was sought. A break was built into the work plan schedule to enable pivoting based on this feedback. One such pivot was the inclusion of subject specific working group offshoots to address specific issues identified by the main CoP.

Another method of sustaining engagement in a CoP is through interest specific and time bound membership. Student active participation in a CoP can also be difficult to gain and even more challenging to sustain. This is due to several factors including competing commitments, imposter syndrome and a lack of 'buy in' to the value of the CoP. Enabling students to join on a casual 'drop in' basis has helped to mitigate against these factors. Students can join sessions during quieter periods of their academic programmes or for topics of interest. They can participate online and take on a passive or active role—with assurances that they will not be 'put on the spot' at any point. It is important to recognise that students come to CoPs to represent themselves and their own experiences rather than as the voice of students. This facilitates a freer flowing communication and negates against positioning or adversary discussions.

In the Hibernia College CoP model, regular opportunities are used for members of the CoP to nominate members of their teams to join while specific issues are being addressed. For example, more members of the Digital Learning and Design team might be brought into the group to contribute to explorations of AI Chat tools. This approach has facilitated a research angle to the CoP. Members can present on areas of research, even those loosely connected to the area of Academic Integrity and seek peer review or research partnerships.

The networking opportunities of a CoP are also integral to the positive experience of members. Put simply, people will engage more actively and willingly in sociable, interesting engagements than they will in prescriptive and detail driven discussions. The optimal space for a CoP is where engagement is relaxed but guided by a clearly understood structure and process. It is possible to facilitate this atmosphere through a rotating chair (thus avoiding one dominant voice), the timetabling of social time or a 'café' structure at the start or end of each session and through deliberate use of and encouragement to use plain English throughout.

The Challenges for Professional Programmes

Professional integrity is generally recognised as fundamental to the quality and assurance of key professions such as teaching and nursing. It involves behaviours, values and attitudes consistent with professional and ethical expectations although there is some debate as to the specific nature of professional integrity. For example, a teleological view of professional integrity is that rules and requests are accepted only if they are sanctioned by a legitimating end and involve making evaluative judgements beyond the reach of the immediate task in hand (Eriksen, 2015). Another view of professional integrity is that of the generic view which holds that it is a virtue that gives reasons to trust professionals as comparable to ordinary integrity as a trait that gives reasons to trust friends (Eriksen, 2015). While an in-depth discussion of professional integrity is beyond the scope of this chapter, it is important to recognise the additional onus on institutions providing qualification of a professional nature, to address integrity challenges in order to prepare future professionals. Guerrero-Dib et al. (2020) describe academic integrity as supporting the development of ethical perspectives which lead to good decision making. Promoting and experiencing academic integrity within the higher education context has a two-fold purpose: to achieve the necessary learnings and skills to appropriately perform a specific profession and to develop an ethical perspective which leads to correct decision making and ethical behaviour in the workplace.

Why Students Cheat

There are a multitude of factors which can contribute to a student either making the decision to cheat or neglecting to engage in good academic practices. These include individual factors, institutional factors, assessment-specific factors, and medium-related factors (Muchemwa, 2023). The 'fraud triangle' (Becker et al., 2006; Holden et al., 2021) proposes three conditions which are present for cheating to occur; opportunity, incentive or pressure and rationalisation or attitude. The collaborative establishment of cultures of integrity through a CoP approach has potential to mitigate against these conditions by establishing policies, expectations, and procedures to decrease opportunities, decrease the incentives to cheat by including academic faculty in embracing more robust assessment practices and promote ethical practices which negate incentives.

However, it should be noted that the establishment of an Academic Integrity CoP has a wider scope than cheating. It is a harnessing of the momentum around academic integrity, provided by the 'big bang' of GenAI, to move away from punitively penalising misconduct (Kaposi & Dell, 2012) and towards a more creative and inclusive response. Furthermore, it represents a rejection of assumptions of moralistic approaches (Bretag et al., 2013) and an acceptance of a need for greater dialogue.

The Vulnerable Student

The broad categorisation of the vulnerable population includes those who experience inequality based on ethnicity, race, gender, sexuality, socio-economic circumstances, or disability (Lauterbach & Dembek, 2023). However, within a higher education context, it refers to a person who, for a variety of reasons, requires extra support to meet the learning outcomes and thus fulfilling their educational pathway. The reasons for this educational vulnerability can overlap with those which distinguish a person as at risk or vulnerable in the wider population but may also include a variety of other reasons such as time paucity, prior educational experiences, performance pressures and lack of context-based understanding of their role as a student in higher education.

Supports within educational institutions for vulnerable students often focus on pastoral aids aimed at increasing the wellbeing of these students through evidence-based health promotions (O'Reilly et al., 2018 or aiding resilience in dealing with adverse situations (Khawaja et al., 2017). However, within the context of academic integrity, vulnerability to nefarious practices or temptations to navigate additional language barriers in written assessment, can lead to second language speaker students also becoming part of this vulnerable population (Pecorari & Petric, 2014). Pecorari (2016) highlights another issue, in that some forms of academic misconduct to which international students might be particularly vulnerable, namely plagiarism, straddle both academic literacy and academic integrity issues. International students who may have never been previously identified as vulnerable, can find themselves in this category due to the additional challenges of working in and being assessed through a second language.

Inadvertent plagiarism or unintentional collusion occur where academic skills are not nurtured and students continue to use skills, practices and learning behaviours which have worked for them in the past (Draper & Fisher, 2020). Cultural differences, academic preparedness and lack of understanding of policies and practices can all play a part in segwaying students into this vulnerable category.

Students can also be vulnerable to the temptations of cheating without actually being classifiable as vulnerable in their profile. Success, the desire for academic attainment and pressures to maintain a high standard, can all lead high performing students to succumb to the pressures of cheating. For this reason, it is important that institutions do not presume to know who is likely to cheat. CoPs with a wide membership including students, can help to negate against overly simplistic interpretations and profiling of a student who is vulnerable to the pressures of assessment.

An advanced stage of a CoP would be when vulnerable students actively, willingly and openly participate. There are a multitude of reasons as to why such a student could be reluctant to join a CoP, primarily based around fears of being exposed, being labelled negatively or highlighting potential poor practices. The 5Ps approach can be utilised to include the aim of developing confidence in the transparency of the CoP, its functioning and the integrity of interactions.

The Hibernia College Academic Integrity Community of Practice

The Hibernia College Academic Integrity CoP was established in April 2023, following on from the immediate controversy surrounding ChatGPT. A key initial aim of the CoP was to move away from a penalty-based system of academic integrity and towards enabling good academic practice, particularly in students who have experienced difficulties with their academic pathways or may fall into the vulnerable category. More specifically, the CoP began with the aim of co-planning pathways to academic integrity with students and for students. A review of existing policies and procedures in this area grounded the initial sessions in the practical task of establishing what we do, how we do it and what we can do better as an institution. The voices of students were recognised as integral to this process and the CoP structure, including the 5Ps approach, facilitated this ongoing communication and collaboration.

Although a CoP facilitates more informal communication, it was considered important to ground the work in a strong theoretical and literature basis. For this reason, a research angle was included, with members working together to conduct literature reviews and write and present research papers on the following areas:

Upholding Academic Integrity

Much of the emphasis in research is currently on how to avoid and mitigate against emerging challenges but the good practices currently in place should be celebrated and encouraged. The CoP focuses on the values of academic integrity (ICAI, 2021) and the potential to learn from and mirror best practice. The CoP acknowledges that good practice is born from the attitudes students bring to their studies. A research output from the CoP has been a project to monitor student attitudes towards academic integrity in the face of GenAI. A longitudinal study capturing changes amongst cohorts of student teachers and nurses in academic integrity in response to GenAI was facilitated through the CoP working groups.

Mitigating Against Academic Misconduct

The area of mitigation is where student input in a CoP can help to ensure mitigations are driven by the needs of students and the challenges they face. Alongside the policy and procedure work to improve investigations of misconduct, the CoP facilitated the development of a research project to learn about the perceptions of pupils in primary classrooms and the student teachers tasked with teaching those pupils how GenAI can be used ethically in the classroom. There is a need for teachers to become acquainted with GenAi in its diverse uses and for pupils to be supported in fostering good and ethical habits from an early age. This research project is currently being facilitated through the CoP structure which has enabled wide participation across different departments.

Detecting Academic Misconduct

Detecting academic misconduct is the area of least comfort in a CoP structure and consequently the most challenging. While a CoP strives to establish a sense of trust and togetherness, the detection of misconduct focuses the conversation on unethical practices, wrongdoing and actions based on the needs of the individual rather than the community. However, a CoP can also move the conversation away from the imposition of penalties and towards acknowledging the needs of a greater good, not only for the individual but for the wider community of the institution and society.

Supporting Vulnerable Students

The CoP is based on the principle of support, enshrined in the reviewed Academic Integrity and Good Practice Policy. This principle holds that all students, regardless of academic standing, are entitled to a respectful investigative process and can hold a reasonable expectation of support from the College to improve their academic practices if found to have engaged in misconduct.

CoP Meetings

The CoP is structured as one body meeting on a 6-week basis, with membership across all college departments and the student body. Between these sessions, meetings of working groups focused on addressing specific areas of interest come together on a more regular basis and report into the main session. The main session is chaired by the Registrar with the invitation for guest chairing and includes invited and volunteered contributions from across the membership. As the CoP has progressed and confidence in its proceedings has grown, volunteer contributions have increased significantly. Prior to each session members are invited to contribute items to the agenda to present themselves or to be presented on their behalf.

Outcomes from the CoP

The CoP has been extremely effective and surprisingly outcomes driven since its inception. This has been facilitated by the 5Ps framework which enables participants to experience the community, offer feedback on these experiences and contribute to its iterative improvement. The working group structure has facilitated this efficiency by allowing small groups to take targeted actions, thus freeing the main group to discuss, consider, review and collaborate.

Policy Development

An early outcome of the CoP was the review of the Academic Integrity and Good Practice Policy. Key changes made to the policy were the inclusion of reference to GenAI, the development of a support structure for students identified as engaging with misconduct and the inclusion of a rating system for levels of misconduct. The policy was also reviewed for the use of plain and accessible English. A new policy on Principles for the Acceptable Use of Generative AI in the Assessment Process was also developed by the CoP and approved by the Academic Board, for the purpose of grading student effort. The policy distinguishes between use of GenAI for research purposes and use of GenAI as a direct response to an assignment. It also sets out the context for the responsible use of GenAI by College Staff, Faculty, Adjunct Faculty and students, with particular consideration of ethical consequences, transparency and fairness in assessment design, development and response.

The Academic Integrity Champions Network

The Academic Integrity Champions Network (AICN) is an innovation which has been piloted in Hibernia College through the work of the CoP. A CoP is a good method for garnering support for new initiatives which may otherwise be seen as being onerous or imposed by an outside force. The overarching purpose of this AICN role is to foster and promote a culture of academic integrity at college level and to facilitate the sharing of resources, learning and good practice at programme level. Furthermore, the Academic Champion (or nominee on their behalf) provides one-to-one support for students found to have engaged in academic misconduct and co-plans a pathway to better and sustainable academic practices. Finally, the Academic Champions work with Peer Champions (students) to discuss issues of relevance with students and provide an opportunity for co-creation of support and resources. The Champions work collaboratively as a network across the College to:

- discuss and propose improvements and developments in Academic Integrity and plan towards Academic Integrity Week;
- consult on the decision-making process in complex or challenging cases of suspected misconduct;
- develop cross-department programme level knowledge and expertise in Academic Integrity;
- consult with the Programme Director on their programme on assessor training and support.

An evolution of the AICN has been the use of case studies of previous investigations into misconduct to promote knowledge sharing and best practice. The AICN members analyse previous investigations for best practice, potential improvements and areas where communications may have hindered the process. They can then feed this growing expertise back into the CoP.

Case Investigation

An area of concern in supporting vulnerable students identified by the CoP was the investigative process when students were brought to the college's attention as suspected of having engaged in misconduct. The occasional nature of the investigator role led to concerns regarding consistency and decision-making processes. The CoP tasked a working group with responsibility for researching and developing the role of the case investigator. Taking advantage of recent NAIN guidelines on the investigative process, this working group developed a comprehensive set of guidelines, Terms of Reference, training materials and samples of good practice. This has escalated into a series of training on the role of the case investigator with a built-in feedback mechanism to develop the role through the CoP.

Conclusion

The success of the CoP can be measured by the sustained willingness on the part of participants to engage with it, since its inception. Institutions embarking on policy and procedure reviews can benefit from such a collaborative approach; however, a key feature of this CoP is that change and evolution are built into the model. The adaptability of the model is an advantage which can be leveraged to support research, enable short term project facilitation and support a nuanced approach to a complex phenomenon. Developing a CoP to address academic integrity is an iterative process. Sessions continue to be well attended with over 80% attendance at most sessions and contributions from individual members in the form of research and presentations are increasing in detail and content. The 5Ps framework has proved effective in ensuring that the CoP has the potential to evolve and foster support along the way. It now functions as a breeding place for innovations in academic integrity and these innovations are filtered through working groups. Going forward, a priority will be the development of the student engagement aspect, particularly in relation to vulnerable students, to ensure that the College continues to be informed by the needs and interests of learners. The CoP is built upon an acknowledgement that community engagement is needed to deal with community challenges. There is an opportunity in the current climate to build on unprecedented interest in academic integrity in order to make significant advances in cultures of integrity in institutions. The CoP approach is emerging as a suitable and sustainable mode to include the voices of all, including vulnerable students, at the table.

References

Adams, S. (2014). *The importance of institutional culture at a technical college.* Dissertations (20090). Paper 337. https://epublications.marquette.edu/dissertations_mu/337/

Becker, D., Connolly, J., Lentz, P., & Morrison, J. (2006). Using the business fraud triangle to predict academic dishonesty among business students. *Academic Education Leadership, 10*(1), 37–52. https://api.semanticscholar.org/CorpusID:221192053

Bretag, T., Mahmud, S., Wallace, M., Walker, R., McGowan, U., East, J., & James, C. (2013). 'Teach us how to do it properly!' An Australian academic integrity student survey. *Studies in Higher Education, 39*(7), 1150–1169. https://doi.org/10.1080/03075079.2013.777406

Campbell, M., & Hourigan, N. (2008). Institutional culture and development education. *Policy and Practice: A Development Education Review, 7*, 35–37. http://hdl.handle.net/10395/2833

Çelik, Ö., & Razi, S. (2023). Facilitators and barriers to creating a culture of academic integrity at secondary schools: An exploratory case study. *International Journal for Educational Integrity, 19*. https://doi.org/10.1007/s40979-023-00125-4

Davidov, E., Schmidt, P., & Schwartz, S. (2008). Bringing values back in: The adequacy of the European social survey to measure values in 20 countries. *Public Opinion Quarterly, 72*(3), 420–445. https://doi.org/10.1093/poq/nfn035

Draper, M., & Fisher, D. (2020). A student and staff partnership in the development of a student directed independent learning toolkit. *International Journal for Students as Partners, 4*(2), 90–96. https://doi.org/10.15173/ijsap.v4i2.4086

Eaton, S. E., Moya Figueroa, B., Butterworth, C., Feledichuk, D., Leslie, K., Lothian, J., Murray, J., Soodeen, C., & Wheatley, B. (2021). Reflections on the first year of integrity hour: An online community of practice for academic integrity. *Canadian Perspectives on Academic Integrity, 4*(1), 4–13. https://doi.org/10.11575/cpai.v4i1.72676

Eriksen, A. (2015). What is professional integrity? *Ekikk I Praksis-Nordic Journal of Applied Ethics, 9*(2), 3–17. https://doi.org/10.5324/eip.v9i2.1836

European Union (2024). *The artificial intelligence act-regulation 2024/1689.* https://eur-lex.europa.eu/eli/reg/2024/1689/oj

Guerrero-Dib, J., Portales, L., & Heredia-Escorza, Y. (2020). Impact of academic integrity on workplace ethical behaviour. *International Journal for Educational Integrity, 16*(2). https://doi.org/10.1007/s40979-020-0051-3

Hackett, S., Kavanagh, Y., Kelly, W., & MacLaren, I. (2024). Enhancing cultures of academic integrity in Irish higher education. In S. Eaton (Ed.), *Second handbook of academic integrity*. Springer International Handbooks of Education. https://doi.org/10.1007/978-3-031-54144-5_172

Holden, O., Norris, M., & Kulmeier, V. (2021). Academic integrity in online assessment: A research review. *Frontiers in Education, 6*. https://doi.org/10.3389/feduc.2021.639814

International Center for Academic Integrity (ICAI). (2021). *The fundamental values of academic integrity, Third Edition*. https://academicintegrity.org/images/pdfs/20019_ICAI-Fundamental-Values_R12.pdf

Kabudi, T., Pappas, I., & Olsen, D. (2021). AI-enabled adaptive learning systems: A systematic mapping of the literature. *Computers and Education: Artificial Intelligence, 2*, 100017. https://doi.org/10.1016/j.caeai.2021.100017

Kaposi, D., & Dell, P. (2012). Discourses of plagiarism: Moralist, proceduralist, developmental and inter-textual approaches. *British Journal of Sociology of Education, 33*(6), 813–830. https://doi.org/10.1080/01425692.2012.686897

Khawaja, N., Ibrahim, O., & Schweitzer, R. (2017). Mental wellbeing of students from refugee and migrant backgrounds: The mediating role of resilience. *School Mental Health, 9*, 2840293–2840293. https://doi.org/10.1007/s12310-017-9215-6

Kier, C., & Ives, C. (2022). Recommendations for a balanced approach to supporting academic integrity: Perspectives from a survey of students, faculty and tutors. *International Journal for Educational Integrity, 18*. https://doi.org/10.1007/s40979-022-00116-x

Lauterbach, M., & Dembek, A. (2023). Gaining insights into the education of vulnerable students: Services in Schools and beyond. *Journal of Research in Childhood Education, 37*(3), 363–365. https://doi.org/10.1080/02568543.2023.2214501

Lave, J., & Wenger, E. (1991). *Situated learning: Legitimate peripheral participation*. Cambridge University Press. https://doi.org/10.1017/CBO9780511815355

Mate, K., & Weidenhofer, J. (2021). Considerations and strategies for effective online assessment with a focus on the biomedical sciences. *FASEB Bio Adv., 9-21*, 9–21. https://doi.org/10.1096/fba.2021-00075

Moriarty, C., & Wilson, B. (2022). Justice and consistency in academic integrity: Philosophical and practical considerations in policy making. *Journal of College and Character, 23*(1), 21–31. https://eric.ed.gov/?id=EJ1332761

Muchemwa, S. (2023). Integrity of university online assessment: Towards developing a function model. *Eureka. Journal of Educational Research, 1*(2), 67–73. https://doi.org/10.56773/ejer.v1i2.8

O'Reilly, M., Svirydzenka, N., & Adams, S. (2018). Review of mental health promotion interventions in schools. *Social Psychiatric Epidemiology, 53*, 647–662. https://doi.org/10.1007/s00127-018-1530-1

Packalen, K., & Rowbotham, K. (2022). Student insight on academic integrity. In S. Eaton & J. Hughes (Eds.), *Academic integrity in Canada. Ethics and integrity in educational contexts, 1*. https://doi.org/10.1007/978-3-030-83255-1_18

Pecorari, D. (2016). Plagiarism, international students, and the second-language writer. Handbook of Academic Integrity, 537-550. doi:https://doi.org/10.1007/978-981-287-098-8_69, Springer Singapore.

Pecorari, D., & Petric, B. (2014). Plagiarism in second-language writing. *Language Teaching, 47*, 269–302. https://doi.org/10.1017/S0261444814000056

QQI, NAIN Generative AI Guidelines for Educators. 2023.: https://www.qqi.ie/sites/default/files/2023-09/NAIN%20Generative%20AI%20Guidelines%20for%20Educators%202023.pdf

QQI, NAIN Academic Integrity Guidelines. (2021). https://www.qqi.ie/sites/default/files/2021-11/academic-integrity-guidelines.pdf

Reedy, A., Janssen, A., & Lockley, A. (2021). A community of practice approach to enhancing academic integrity policy translation: A case study. *International Journal for Educational Integrity, 17*. https://doi.org/10.1007/s40979-021-00080-y

Tatum, H. (2022). Honor codes and academic integrity: Three decades of research. *Journal of College and Character, 23*, 32–47. https://doi.org/10.1080/2194587X.2021.2017977

Tikhonova, E., & Raitskaya, L. (2023). Academic integrity: Author-related and journal related issues. *Journal of Language and Education, 9*(3), 5–10. https://doi.org/10.17323/jle.2023.18489

Vaughan, G. B. (1994). In I. A. M. Cohen, F. B. Brawer, & Associates. (Eds.), *Managing community colleges*. Jossey-Bass, Inc.

Chapter 7
Research Camp: Building Cultures of Integrity with Peer Influence

Bradford Barnhardt, Asha Torczon, John Robertson, Rahima Thomas, Anusuya Panchagatti, and Vandana Sharma-Ferguson

Abstract This chapter introduces "Research Camp", detailing its foundation in academic integrity literature, its design, and its relevance in the era of generative AI. It also presents evidence of Research Camp's effectiveness in building a culture of integrity at Woodstock School in India. Conducted six times since 2021, Research Camp lasts 1–2 days and features a series of workshops and engagements aimed at helping faculty and students hone basic research skills (paraphrasing, citing, referencing, source evaluation, and using an online research management platform). Peer interaction is central to the Research Camp experience. Teachers and students engage with peers in large- and small-group sessions, and one-on-one mentorship among students enhances skill-building and provides a powerful opportunity for the transmission of peer norms. Student mentors, who have passed Research Camp in the past and received digital badges, revisit their research skills and guide mentees in the completion of research mini projects, which must be submitted by the end of the event. The overarching aim of Research Camp is to give rise to virtuous cycles of peer influence and behaviour that build and sustain an institutional culture of academic integrity. Data analysis presented in this chapter shows that since the introduction of Research Camp, cheating has decreased, and students' perceptions of their peers' academic integrity have improved. This suggests a cultural shift in favour of academic integrity.

B. Barnhardt (✉) · R. Thomas · A. Panchagatti · V. Sharma-Ferguson
Woodstock School, Mussoorie, India
e-mail: Principal@kis.in; RahimaThomas@woodstock.ac.in; AnusuyaVijay@woodstock.ac.in; vandana.ferguson@sais.edu.sg

A. Torczon
Cambridge University, Cambridgeshire, UK
e-mail: amt92@cam.ac.uk

J. Robertson
Hanoi International School, Hanoi, Vietnam
e-mail: JRobertson@hisvietnam.com

M. Davis, C. Baptista (eds.), *Ethics and Integrity in Education (Practice)*, Ethics and Integrity in Educational Contexts 10,
https://doi.org/10.1007/978-3-031-95686-7_7

Keywords Cultures of academic integrity · Peer influence · Social comparison theory · Generative AI

Introduction

How can educational institutions build cultures of integrity? Since January 2021, Woodstock School, a K-12 international boarding school in India, has tackled this question by leveraging peer influence through an initiative dubbed "Research Camp". Peer influence is both a prominent variable in published research on academic integrity and a key mechanism for the transmission of knowledge and behaviour that shape culture (Bailey & Peoples, 1998). While traditional research ethics programs focus principally on teaching skills and promoting certain attitudes (e.g., Bebeau & Thoma, 1999; Evans & Craig, 1990), Research Camp emphasises the *importance* of these skills and attitudes through peer interaction and peer mentoring.

Building cultures of integrity has become much more challenging since the emergence of generative AI (GenAI), which makes detecting academic misconduct difficult or even impossible (Plata et al., 2023). By offering a free, low-risk temptation to plagiarise, GenAI does more to encourage cheating behaviour than perhaps any other invention in history. Misuse of GenAI by growing numbers of students worldwide threatens to undermine cultural norms of academic integrity built by generations of teachers and school leaders. Rebuilding these cultures is an urgent need.

Since the 1990s, scholars have advocated for the importance of building cultures of academic integrity (e.g., Alschuler & Blimling, 1995), noting that peer norms can irresistibly affect students' integrity behaviours (Bertram Gallant et al., 2014). Prominent academic integrity researcher Donald McCabe, a leading advocate for cultures of integrity, argued that schools and universities should involve student leadership and faculty engagement in "ongoing dialogue about academic integrity that begins with recruiting, continues in orientation sessions and initiation ceremonies, and continues throughout the program" (McCabe et al., 2006, p. 302; see also McCabe, 2005; McCabe & Katz, 2009). Research Camp answers this call, and we hope to inspire other schools and universities to do the same.

Bailey and Peoples (1998, p. 23) define *culture* as "the socially transmitted knowledge and behaviour shared by some group of people". Their emphasis on *social transmission* implies that building cultures of integrity requires social interactions focused on academic integrity. Many researchers have documented the role of social transmission in propagating integrity-related behaviours, usually highlighting its corruptive tendency to feed vicious cycles where students, recognizing that their peers cheat, accept this as a norm and cheat more themselves. However, published research has barely considered whether this feedback loop is reversible, that is, whether peer influence can promote virtuous cycles that foster positive cultures of academic integrity.

Research Camp helps participants build traditional research skills and, in keeping with Bailey and Peoples's (1998) definition of culture, creates a context in which students socially transmit the importance of academic integrity. We theorise that when students perceive norms of academic integrity among their peers, they tend to emulate those behaviours. During Research Camp, returning students who have achieved the rank of "Research Mentor" in prior Research Camps help new students with assignments, transmitting in the process both the knowledge and the importance of research ethics. Research mentors also influence each other by working together in teams. Simultaneously, returning students who are not yet Research Mentors review basic research skills in small group sessions, while teachers work together to integrate research skills into unit plans.

This chapter explores Research Camp as a means of building cultures of academic integrity in the GenAI era. We begin by reviewing key literature on peer influence and introducing Social Comparison Theory and the Coordination Effect as lenses for understanding how peer influence can lead to social contagions. We then provide the rationale for Research Camp and describe its design, including an overview of its broader ecosystem of support at Woodstock School. In the third section, we present data indicating that cheating behaviour has decreased, and students' perceptions of their classmates' academic integrity have improved at Woodstock since the introduction of Research Camp in 2021. Finally, we discuss our findings, review limitations and future directions, and summarise our arguments in the conclusion.

Literature Review

This section reviews key literature linking peer influence with cheating behaviour and contagion effects over the last century. It also highlights the "coordination effect", first proposed by Arthur (1988), and a model of social feedback loops that Festinger (1954) described in his well-known formulation of social comparison theory.

The Role of Peer Influence in Integrity Behaviours

Nearly 100 years of research across secondary and tertiary settings suggests that peer influence drives academically dishonest behaviours (e.g., Bin-Nashwan et al., 2023; Bowers, 1964; Eisenberg, 2004; Galloway, 2012; Hartshorne & May, 1928; Latova & Latov, 2008; Malesky et al., 2022; Nora & Zhang, 2010; Teodorescu & Andrei, 2009; Zhao et al., 2022). While research on prosocial peer influence does exist (e.g., Brechwald & Prinstein, 2011; Luo et al., 2023), it has crossed over minimally or not at all into research on cheating. Academic integrity researchers have occasionally implied that prosocial peer influence could help foster institutional cultures of integrity (e.g., McCabe, 2005), but no study reviewed for this chapter reported on how to take advantage of that possibility. Hence, this work addresses a significant gap in the literature.

In his influential book *Student Dishonesty and its Control in College*, Bowers (1964, p. 196) observed that "the level of disapproval of cheating among the student's college peers" was the highest determining factor in altering cheating practices at both high school and college levels. More recently, Zhao et al. (2022, p. 13) found in a meta-analysis on academic dishonesty, peer-cheating and culture, that perceived peer cheating was one of the strongest drivers of individual cheating behaviour and "significantly stronger than factors such as age, gender, conscientiousness, and achievement motivation". Whitley (1998) and Whitley et al. (1999) reported similar findings in two earlier meta-analyses.

Researchers have studied variables in relation to academic dishonesty across three broad groups: personological, situational, and emergent. Although there is some overlap between these categories, they reflect the historical tendency to attribute academic dishonesty to individual characteristics, learning situations, or psychological states induced by combinations of individual and contextual factors. While researchers typically treat peer influence as a situational variable, significant evidence suggests that it plays a special, mediating role between contextual and emergent variables. For instance, Murdock et al. (2004) found that students co-constructed their perceptions of teachers, which affected whether they felt cheating was justifiable. Murdock et al. (2008) found that students blamed the teacher's pedagogy for cheating at both the individual and aggregate levels. Similarly, in a longitudinal modelling study, Barnhardt (2014) found that peer norms completely mediated the effect of perceived teacher quality on whether students viewed cheating as justifiable.

Social Comparison Theory

Social comparison theory (SCT) (Festinger, 1954) offers a framework that numerous researchers have used for understanding how peer influence mediates between class context and cheating behaviour (e.g., Bäker & Mechtel, 2019; Broeckelman-Post, 2008; Koul et al., 2009; Nora & Zhang, 2010; Zanetti & Butera, 2023). SCT posits that individuals formulate and validate their beliefs in a manner that is consistent with their social peers. By this view, peers co-construct beliefs about classes, teachers, learning experiences, and schools that affect their integrity behaviours (Murdock et al., 2004). Students may co-construct such beliefs openly or tacitly. They might speak to each other about how they perceive, like, or dislike their teachers or classes, or they might pick-up on peers' social-emotional cues such as body language and behaviours. Research suggests that a student's belief that their classmates are cheating successfully strongly influences whether they also decide to cheat (e.g., Hartshorne & May, 1928; Zhao et al., 2022). Without intervention, such dynamics can lead to vicious cycles wherein growing numbers of students perceive

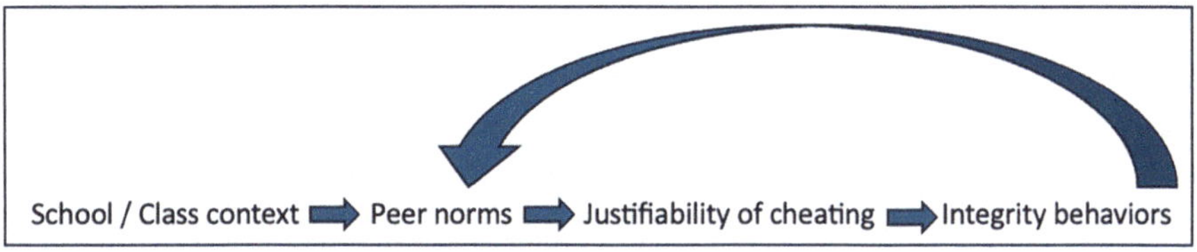

Fig. 7.1 Feedback loop implied by social comparison theory

that cheating aligns with peer norms, feel it is justified for that reason, and engage in it themselves (please see Fig. 7.1).

Social Contagion

Researchers have empirically observed the feedback loop shown in Fig. 7.1. When this loop occurs on a large scale, it is commonly referred to as a *contagion* (e.g., Baird, 1980; Carrell et al., 2008; Woo et al., 2018). In 1966, Walker et al. conducted an experimental study involving a series of three tests in a college class. They designed the tests such that cheating could be easily detected, with the professor agreeing to ignore it. Results showed that 23% of students cheated on the first test, increasing to 64% and 86% on the second and third tests. Those who cheated successfully on the first test thus appeared to have shared this information with classmates, who then also cheated. Similarly, qualitative studies at the high school level have found that students often inform peers about how easy it is to cheat on assessments or in their teachers' classes (Schraw et al., 2007).

The peer dynamics that drive cheating contagions exemplify what Arthur (1988) called the "coordination effect", whereby "the more consistently a norm is observed in society, the greater the costs incurred by an individual deviating from it" (Magnus et al., 2002, p. 131). In other words, the social costs of cheating diminish as peer norms shift from valuing academic integrity to justifying academic dishonesty. However, this raises the question of whether contagion effects ever work in prosocial directions, creating virtuous feedback loops instead of vicious ones.

Research suggests that prosocial contagions are indeed possible. Tsvetkova and Macy (2014, p. 1) found, for instance, that people who receive help from others become more helpful themselves, which can lead to a "contagion of generosity". Similarly, Barry and Wentzel (2006) found that adolescents were more likely to behave in prosocial ways if they perceived their friends doing so. Such findings suggest that the dynamic shown in Fig. 7.1 can include either honest or dishonest behaviours. As Tsai (2012, p. 154) argued, the most effective educators are often peer leaders; "a school can train these student leaders to foster academic integrity in class".

Rationale for Research Camp

The initial motivation behind Research Camp was to empower students with the knowledge and skills necessary to conduct research with integrity. This aim evolved into building a culture of integrity at Woodstock on the basis suggested by SCT, whereby individuals shape their own attitudes and behaviours by observing those of their peers. By creating opportunities for positive peer influence related to academic integrity, Research Camp aspires to foster virtuous feedback loops that reinforce it as a norm.

Design of Research Camp

This section provides an overview of the design and operation of Research Camp, focusing on how it brings participants together for peer interactions. We first explain how Research Camp is situated within a broader ecosystem of support for academic integrity at Woodstock School. We then explain how it is organised as an event, including its master objectives and logistics.

Academic Integrity Ecosystem

Research Camp emerged in 2022 within a loosely organised ecosystem of support for academic integrity (please see Fig. 7.2). This ecosystem included a student-led academic honesty council (AHC) that works with school leadership (SL) to steward the school's Academic Honesty Policy, a chapter of the National Honor Society (NHS) that represents academic excellence and works with the AHC on programs that support high standards for integrity among students and staff, a library that conducts research skills classes, teachers who teach research skills, and the International Baccalaureate which sets strict standards for academic integrity on internal and external assessments. A policy on the use of GenAI in classrooms is a recent addition to this ecosystem. The introduction of Research Camp provided this ecosystem with a galvanising, scheduled event that, since its inception, has graduated over 200 student research mentors who uphold academic integrity among their peers.

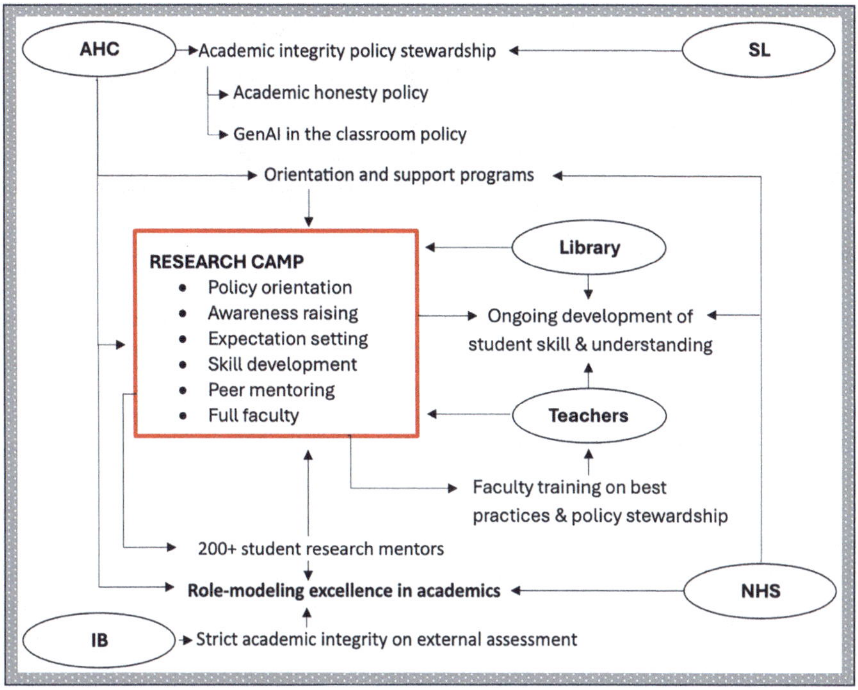

Fig. 7.2 Ecosystem of support for academic integrity. *Note. AHC* Academic Honesty Council; *SL* School Leadership; *NHS* National Honor Society; *IB* International Baccalaureate

Organization of Research Camp as an Event

This section provides an overview of a "typical" Research Camp. Over the course of six iterations, Woodstock has experimented with different versions of Research Camp, varying in duration and organisation. The version presented below is representative because its components and flow are current, have been found to work well, and/or have remained consistent over time. We first outline three master objectives (1. skills development, 2. social engagement, and 3. sustainability) and detail the logistical elements designed to achieve these objectives.

Master Objectives of Research Camp

Skills Development Students and teachers focus on honing their research integrity skills. Students receive direct instruction on fundamental skills (paraphrasing, citing & referencing, source evaluation, and using Woodstock's online research management platform) in group workshops. New mentees (who are new to Woodstock School) practise these skills by completing research mini projects under the

Fig. 7.3 Research mentor badges

guidance of their research mentors. Prior to helping their mentee, student mentors review basic research skills in large group sessions and take an assessment to ensure they have mastered the skills they need to teach. Returning mentees complete a series of small-group workshops, after which they either pass a skills assessment or complete a mini project. Separately, teachers review academic policy and practice and work together to integrate research skills into taught units.

Social Engagement Social engagement is what makes Research Camp activities effective. New mentees begin each day with a selection of ice-breaker activities followed by a series of interactive workshops, pair work on mini projects, and then one-on-one collaboration with their assigned research mentors, who help them complete their individual research mini projects. New mentees are typically the most excited, open, and ready to learn, so fostering positive relationships during Research Camp—with each other and their mentors—is crucial.

Student mentors are organised into teams, where each team is led by at least one mentor who has achieved Level 2 or above (please see Fig. 7.3). Mentor teams support each other as they guide mentees through their mini projects. Mentor teams later internally moderate assessments of mentees' mini projects (not of their own mentees). Mentor-mentee engagements are crucial for establishing that Woodstock students hold academic integrity in high esteem.

Returning mentees, those returning students who have not yet passed Research Camp, rotate through workshops in small groups where they also engage in ice-breaker activities and interact directly with each other and workshop leaders. Teachers work in departmental groups to develop or hone research-based units, followed by sharing activities across departments.

Sustainability The sustainability of Research Camp is a high priority in two ways: ensuring the experience has a lasting effect on participants and allowing the program to grow year-over-year. Students who pass Research Camp become mentors and receive digital mentor badges (please see Fig. 7.3). Approximately 250 students participated in the most recent Research Camp with 90 mentees receiving one-on-one mentoring from a student mentor. High numbers of student mentors are needed to sustain one-to-one mentor-mentee pairings. Student mentors also help sustain the spirit of Research Camp over time by proudly displaying their digital mentor badges in their email signatures.

The work teachers do during Research Camp is sustained in the research-based units they develop, deliver, and refine over time. In these ways, Woodstock has sustained Research Camp over seven semesters, equipping students with increasingly higher skills and knowledge for conducting research ethically and independently.

Logistical Design

Research Camp typically lasts one or two days and includes students and teachers in middle and high school. It has evolved to include four separate tracks: (1) student research mentors, (2) mentees, (3) returning mentees, and (4) teachers. A common component across all tracks is the review of Woodstock's policies on academic integrity and GenAI. The details of each track are provided below.

Track 1: Student Research Mentors Students become student research mentors (hereafter *mentors*) after passing at least one Research Camp. As a privilege, mentors begin later in the morning than others (usually about 10 am). Their first task is to review the basics of research integrity and take an online assessment. The assessment results indicate their readiness to guide mentees effectively. By performing well, mentors qualify for new digital badges (please see Fig. 7.3). They then meet with their assigned mentees and help them navigate their mini-report assignments with integrity. Once their mentees submit their mini projects, research mentors become assessors. To avoid conflicts of interest, mentors do not evaluate submissions from their own mentees. Each mentor assesses one mini report, provides feedback according to a rubric, and gives the new mentee 24 h to correct and resubmit. Finally, mentors meet in small groups supervised by teachers to moderate their assessment results.

Track 2: New Mentees New mentees are new students who receive one-on-one mentoring from elder classmates during Research Camp. Following welcome remarks, ice-breaker activities, mentees move through a series of four workshops covering basic research integrity skills (paraphrasing, citing and referencing, source evaluation, and using an online research management platform). Each mentee receives a work packet with the following: an individualised mini-report prompt, instructions with hyperlinked tutorial videos, and an assessment rubric with hyperlinked worked examples. Individualised report prompts prevent collusion, following the 'design-it-out' principles of plagiarism prevention (Heckler et al., 2013). Mentees work independently on their mini reports for about an hour before their mentors join them to review their progress and provide guidance. Mentees submit finished mini reports for rapid marking at the end of Research Camp. They receive rubric-based feedback from their assigned assessor the next day and have 24 h to correct and resubmit. They must satisfy all aspects of the rubric to pass and become research mentors.

Track 3: Returning Mentees Approximately 45% of students pass Research Camp each year. Returning mentees are students who have not yet passed Research Camp. These students break into small groups and proceed through four workshops covering the above-mentioned research integrity skills. After their workshops, they complete an online skills assessment. Those who pass receive their first digital badge, while those who do not pass the assessment receive work packets similar to new mentees. They have approximately 2 h to complete their mini reports, with workshop leaders (who may be teachers or research mentors) circulating to provide support. At the end of the 2 h, they submit mini reports to teachers for final marking.

Track 4: Teachers Teachers begin Research Camp in a large-group session and break into departmental groups to develop or refine units of study for the coming year that incorporate the fundamental skills of research integrity. Once most teachers reach a stopping point, they pair-and-share their units with colleagues outside their department. Volunteers then share their units with the whole group. The day wraps up with a fun, online trivia-style quiz covering skills and facts related to research integrity policy and practice. Finally, a week or so later, they meet with small groups of student mentors to supervise moderation of mini reports.

This section has presented an overview of the organisation of a typical Research Camp, focusing on three master objectives and four 'tracks' for participant groups. Research Camp is a recent addition to a broader ecosystem that includes traditional forms of support for academic integrity, such as an academic honesty policy, an academic honesty council, and various leadership roles for students and staff. GenAI undermines many traditional forms of support for academic integrity in schools (Plata et al., 2023), highlighting the need for innovative means to build and sustain cultures of academic integrity. The next section examines data on how and to what extent two key components of culture—perceived peer norms and behaviour—have changed since Woodstock introduced Research Camp in 2021.

Data Collection

This section examines two types of data relevant to Bailey and Peoples's (1988) definition of culture, namely (1) longitudinal survey data on students' perceptions of their classmates' academic integrity, i.e., transmitted knowledge, and, by extension (2) data on the pervasiveness of cheating behaviour from the AHC's hearing records over 5 years. By this definition, improvements in these variables since the introduction of Research Camp suggests that Woodstock's culture of academic integrity has strengthened.

Participants

All participants in this study were students in grades 6–10, including the groups described above as *new mentees* and *mentors. New mentees* tend to be younger on average than *mentors* since they have just started at Woodstock, whereas *mentors* have been at Woodstock for at least 1 year. As Table 7.1 shows, *new mentees* include a higher concentration of grade 6 students, whereas *mentors* include a higher concentration of grade 10 students.

Results

This section focuses on data from student surveys and AHC hearing records. First, it presents percentage differences and *t*-tests of student survey data indicating whether students think their classmates take honesty on class assignments seriously and whether honesty on class assignments is an important value at Woodstock. Second, it presents data from the past 5 years of AHC records, charted to illustrate trends in student behaviour. Data from AHC records serve as proxies for the prevalence of cheating behaviour among Woodstock students before and after the introduction of Research Camp.

Student Survey Data

An anonymous survey was used in the spring and fall of 2023 to collect student feedback with Likert scaled items (Strongly agree, Agree, Somewhat agree, Disagree, Strongly disagree). The composition of the groups surveyed changed semester-over-semester, especially as ninety new students joined in fall 2023. The analyses below focus on two items that gauge how students regard their peers'

Table 7.1 Grade level distribution of survey respondents

	Spring 2023					
	Grade 6	Grade 7	Grade 8	Grade 9	Grade 10	Total
Mentees	16	15	20	11	0	62
Mentors	0	5	14	22	5	46
Whole group	16	20	34	33	5	108
	Fall 2023					
	Grade 6	Grade 7	Grade 8	Grade 9	Grade 10	Total
Mentees	18	22	17	5	4	66
Mentors	0	7	11	12	28	58
Whole group	18	29	28	17	32	124

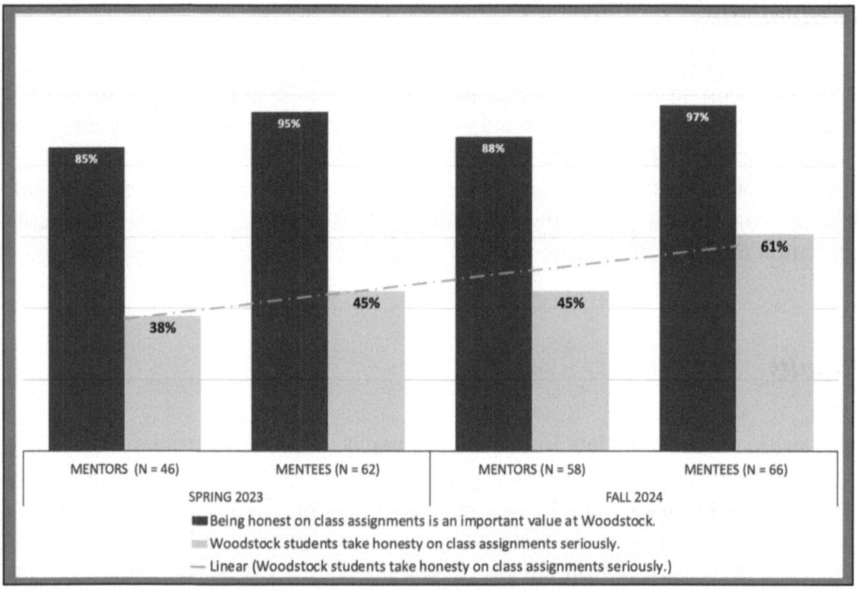

Fig. 7.4 Semester-Over-Semester Comparison: Perceptions of whether academic integrity honesty is an important value and whether students take it seriously

behaviours and how they regard the importance of academic honesty as a value at Woodstock.

Percentage Differences

Figure 7.4 presents differences in the percentage of students who either "strongly agreed" or 'agreed' with two key survey items. Agreement with the first statement, "Being honest on class assignments is an important value at Woodstock" (hereafter "honesty is an important value"), remained stable between spring and fall. Agreement with this item rose just 3% among *mentees* and 2% among *mentors*. By contrast, agreement with "Woodstock Students take honesty on class assignments seriously" (hereafter "WS students take honesty seriously") increased more substantially, with *mentees* showing an increase of 16% increase and *mentors* of 7%. *T*-tests were used to examine the significance of these mean differences.

Mean Differences

This section presents *t*-tests of mean differences corresponding to the survey items reflected in Fig. 7.4. We used the *Social Science Statistics* online *t*-test calculator (Socscistatistics, n.d.) to calculate all *t*-tests. Results show that *mentees* agreed more strongly than *mentors* that honesty is important to their peers and an important value

Table 7.2 Means and standard deviations of responses to integrity-related survey items, Spring and Fall Semesters, 2023

	Honesty is an important value						WS students take honesty seriously					
	Spring 2023			Fall 2023			Spring 2023			Fall 2023		
	N	*M*	*SD*	*N*	*M*	*SD*	*N*	*M*	*SD*	*N*	*M*	*SD*
Mentees	62	4.53	.82	66	4.82	.43	62	3.52	1.07	66	3.97	.89
Mentors	46	4.33	.76	58	4.36	.77	46	3.30	1.07	58	3.36	1.09

at Woodstock. Additionally, their level of agreement with both items increased from Time 1 (spring) to Time 2 (fall).

Within-Times T-Tests

Mean differences between mentors and new mentees in the spring were non-significant, whereas in the fall, new mentees indicated higher average agreement than mentors on both "honesty is an important value" (one-tailed $t = 4.17$, $p < .01$) and "WS students take honesty seriously" (one-tailed $t = 3.41$, $p < .01$). Please see Table 7.2 for group means and standard deviations.

Between-Times T-Tests

Mentees showed higher average agreement in the fall than the spring on both items, 'honesty is an important value' (one-tailed $t = 2.49$, $p < .01$) and "WS students take honesty seriously" (one-tailed $t = 2.61$, $p < .01$). Between-times mean differences for *mentors* were non-significant.

Academic Honesty Council Hearing Records

The AHC recorded the data presented below over the period Fall 2019—Spring 2024. The AHC is a student-led body that addresses incidents of academic dishonesty and aims to educate teachers and students on academic integrity. When a teacher raises concern that a student may have engaged in academic dishonesty, the AHC reviews the evidence and holds a hearing with the student. AHC hearing records capture how many students attend investigative hearings and for what infractions. They also capture the students' grade levels, the investigative findings, and the consequences assigned. These records provide insight into trends of academic dishonesty over the years. While much cheating goes undetected, we assume that the proportion of cases caught by teachers remains reasonably stable over time, such that AHC hearing records serve as a proxy measure of actual cheating in the student body.

Incidence of Academic Honesty Infractions Heard by the AHC

Figure 7.5 shows that the prevalence of academic dishonesty at Woodstock has decreased since 2021. Academic dishonesty reached a peak in 2020–2021, the only academic year when classes at Woodstock were entirely online. Academic years that involved some months of online learning, 2019–2020 and 2021–2022, had the second and third highest incidences of academic dishonesty, respectively. However, an earlier AHC record, from 2017–18 captured 31 cases, indicating that the incidence of academic dishonesty hearings between 2019 and 2022 was roughly consistent with past years. Unfortunately, data from 2018–19 could not be located for this study.

During the 2021–22 academic year, the number of academic honesty infractions reported to AHC dropped from 36 in the prior year to 30 (a 17% decline). The number of AHC hearings fell again during 2022–23, from 30 to 17 (a 43% decline), and again during 2023–24, from 17 to 14 (an 18% decline). Research Camps in each of the latter years were held during the first week of the first semester.

Overall, academic dishonesty was more common on take-home assignments (69.2%) than in-class tests (30.8%). All academic dishonesty on take-home assignments involved plagiarism; however, it is interesting that the AHC has recorded just three cases of plagiarism involving GenAI, ever (all during the 2023–24 academic year).

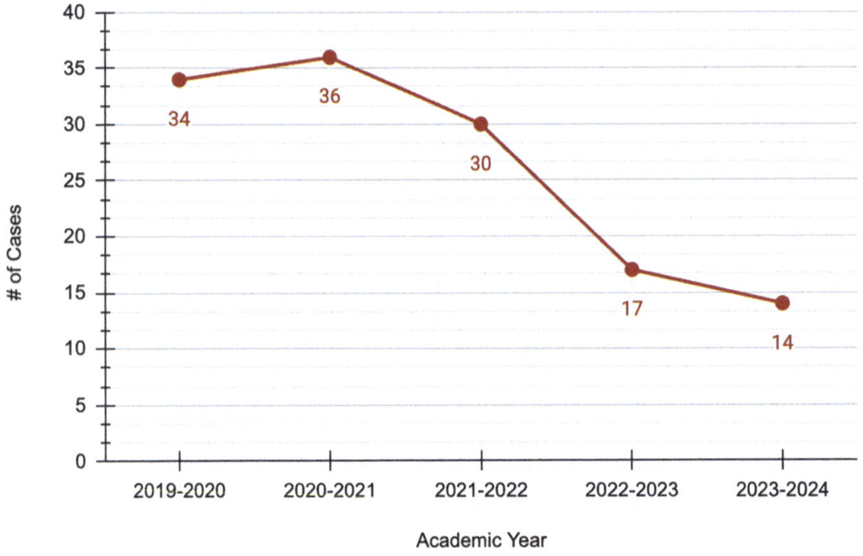

Fig. 7.5 Incidence of academic dishonesty infractions reported to Woodstock's AHC, 2019–2024

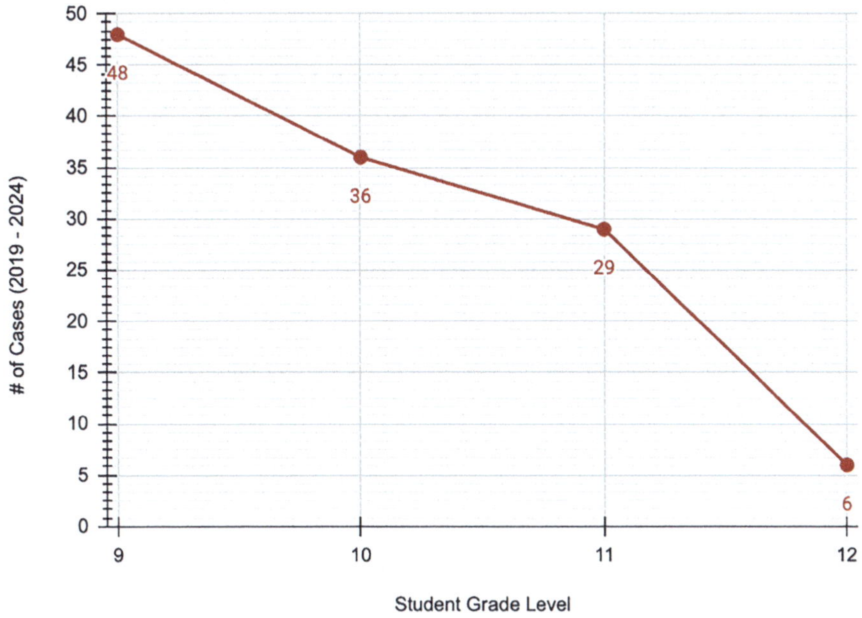

Fig. 7.6 Incidence by grade-level of academic dishonesty infractions reported to Woodstock's AHC, 2019–2024

Incidence of Academic Honesty Infractions by Grade Level

AHC records indicate that the prevalence of academic dishonesty at Woodstock decreases as students progress to higher grade levels (please see Fig. 7.6). Over the past 5 years, the highest incidence of academic dishonesty has been observed among Grade 9 students. Grade 9 students often struggle with the increased academic pressure of their first year of high school; they are also often less aware of the consequences academic dishonesty can have in college application processes. By contrast, the AHC has recorded just six cases of academic dishonesty among Grade 12 students over the last 5 years.

Discussion

The results of this study indicate that the culture of integrity at Woodstock is improving both in terms of students' perceptions of peer norms and their actual behaviour. These results suggest that perceptions and behaviours are shifting together, in the manner predicted by SCT (please see Fig. 7.1). The theorised catalyst for this shift is peer influence that (1) is initiated during Research Camp (2) has the greatest effect on new mentees, and (3) is sustained throughout the academic year by peer mentors' representations of academic integrity, such as including their digital badges in their email signatures.

The percentage of *new mentees* who agreed with both "honesty is an important value" and "WS students take honesty seriously" increased from Time 1 to Time 2, and *t*-tests showed that the level of increase was significant. As of Time 2, *new mentees* also agreed significantly more than *mentors* on both items. These findings are consistent with the expectation that *mentors* exerted positive influence on *new mentees*, lending credibility to the hypothesis that peer influence during Research Camp promotes the perception among *new mentees* that academic integrity is an important norm at Woodstock.

Reports of cheating to the AHC decreased by 17%, 43%, and 18% over three consecutive years following the introduction of Research Camp. This resembles the increase in cheating behaviour among students in a college class, from 23% to 64%, to 86% (increases of 23%, 41%, and 22%, respectively), that Walker et al. (1966) described as a "contagion". From a coordination effect point of view, a coalescence of peer norms favourable to academic integrity should affect behavioural outcomes by increasing the social costs of academic dishonesty. An institution-wide decrease in cheating behaviour of 61% since the introduction of Research Camp supports this possibility. Additionally, the number of hearings consistently fell as students progress to higher grade levels, gain in maturity and, on average, have more years of experience at Woodstock.

AHC hearings involving GenAI were rare, which is surprising. It would be naive to suggest that students are not resorting to the unethical use of AI at Woodstock; it is more likely that current methods of AI plagiarism detection are insufficient. However, a significant drop in the total number of AHC hearings began at least 18 months before ChatGPT's release in November 2022, suggesting a real decrease in cheating behaviour rather than a shift to GenAI plagiarism.

Limitations

Research Camp was not originally conceived or designed as an opportunity to do publishable research, so this study lacks some of the formality of true experimental or quasi-experimental design. As a result, the study faced several important limitations, foremost among them being the lack of a testable statistical model, a control group, a diverse sample (as it came from a single school where the researchers and designers of Research Camp work as educators or study as students), and small sample sizes of survey respondent groups ($N = 46$–66). Moreover, the student survey data noted in this study came from single-item measures. Although single-item measures for psychometric research have gained in acceptability over recent decades (Allen et al., 2022), using multiple-item measures for key constructs would improve generalizability.

While we feel that the survey and behavioural data presented in this chapter make a convincing case for Research Camp, it should be considered preliminary.

Many potential confounding variables in the broader support ecosystem at Woodstock call these results into question. For instance, changes in leadership personnel within various components of the support ecosystem shown in Fig. 7.2 could have directly or indirectly affected academic integrity behaviours and attitudes.

Future Directions

A high-priority future direction is to have more schools and universities run Research Camps and share data, innovation, and insight. GenAI is a global phenomenon; therefore, our push to resist its negative influences must also be global. Replicating Research Camp at multiple institutions would create a more diverse and generalizable dataset for testing statistical models, supporting iterative improvements, and potentially giving rise to new and better ideas for building cultures of academic integrity.

Participating institutions should use a variety of approaches to measure impact, including experimental methods, quasi-experimental methods, and high-quality psychometric surveys to test theoretically grounded statistical models that express key hypotheses including mediating and moderating variables. Considering evidence that peer norms mediate the effects of learning context on cheating behaviours and attitudes (Barnhardt, 2014; Murdock et al., 2004), models that measure antecedents of perceived peer norms could explain more variance than those focused solely on peer norms and behavioural outcomes. Broader models would also enable more sophisticated analytical methods such as structural equation modelling.

Another important future direction is the continued evolution of Research Camp. The model used at Woodstock has changed considerably over six iterations. It has become more student-centred, thanks to an increasing number of student research mentors who can work with classmates. Currently, approximately 55% of Woodstock students are research mentors, enabling a 1-to-1 ratio with mentees. Key considerations for future Research Camps include devising models that are more adaptable to specific needs and better integrated into ongoing curricula, while retaining the centrality of peer interactions. Future Research Camps should include workshops and interactions designed to guide students on the appropriate use of GenAI. Ultimately, programs involving Research Camp models should aim to fully integrate research ethics into classroom practice.

Conclusion

GenAI presents a potent threat to schools' cultures of integrity. Many scholars have emphasised these cultures as a force that can go "beyond deterrence, detection, and punishment" (Bretag, 2013, p. 2). This chapter presented Research Camp as a program whereby schools can build and sustain cultures of integrity. Returning to

Bailey and Peoples's (1998) definition of culture as "the socially transmitted knowledge and behaviour shared by some group of people," this study suggests that a concerted shift in perceived peer norms (a type of knowledge) and integrity-related behaviour has taken place at Woodstock School, amounting to a shift in its culture of academic integrity. Peer influence is the theorised catalyst for this shift, and Research Camp acts as a crucible of peer influence. Together, these observations support the belief that Research Camp can help meet the pressing need of all educational institutions to build and maintain cultures of integrity in the era of GenAI.

References

Allen, M. S., Iliescu, D., & Greiff, S. (2022). Single item measures in psychological science. *European Journal of Psychological Assessment, 38I*(1), 1–5. https://doi.org/10.1027/1015-5759/a000699

Alschuler, A. S., & Blimling, G. S. (1995). Curbing epidemic cheating through systemic change. *College Teaching, 43*, 123–125. https://doi.org/10.1080/87567555.1995.9925531

Arthur, W. B. (1988). Self-reinforcing mechanisms in economics. In P. W. Anderson, K. J. Arrow, & D. Pines (Eds.), *The economy as an evolving complex system* (pp. 10–31). Addison-Wesley.

Bailey, G. A., & Peoples, J. G. (1998). *Introduction to cultural anthropology*. Wadsworth Publishing.

Baird, J. S. (1980). Current trends in college cheating. *Psychology in the Schools, 17*, 515–522. https://doi.org/10.1002/1520-6807(198010)17:4<515::AID-PITS2310170417>3.0.CO;2-3

Bäker, A., & Mechtel, M. (2019). The impact of peer presence on cheating. *Economic Inquiry, 57*(2), 792–812. https://doi.org/10.1111/ecin.12760

Barnhardt, B. (2014). *A social contract model of 'disintegrity' within the dual-process paradigm of moral psychology: Reducing the scope of the 'belief-behavior incongruity'* (Doctoral dissertation, University of Sydney.) http://hdl.handle.net/2123/13895

Barry, C. M., & Wentzel, K. R. (2006). Friend influence on prosocial behavior: The role of motivational factors and friendship characteristics. *Developmental Psychology, 42*(1), 153–163. https://doi.org/10.1037/0012-1649.42.1.153

Bebeau, M. J., & Thoma, S. J. (1999). "Intermediate" conceptions and the connection to moral education. *Educational Psychology Review, 11*, 343–360. https://doi.org/10.1023/A:1022057316180

Bertram Gallant, T., Van Den Einde, L., Ouellette, S., & Lee, S. (2014). A systemic analysis of cheating in an undergraduate engineering mechanics course. *Science and Engineering Ethics, 20*(1), 277–298. https://doi.org/10.1007/s11948-013-9435-6

Bin-Nashwan, S. A., Sadallah, M., & Bouteraa, M. (2023). Use of ChatGPT in academia: Academic integrity hangs in the balance. *Technology in Society, 75*, 102370. https://doi.org/10.1016/j.techsoc.2023.102370

Bowers, W. J. (1964). *Student dishonesty and its control in college*. Bureau of Applied Research, Columbia University.

Brechwald, W. A., & Prinstein, M. J. (2011). Beyond homophily: A decade of advances in understanding peer influence processes. *Journal of Research on Adolescence, 21*(1), 166–179. https://doi.org/10.1111/j.1532-7795.2010.00721.x

Bretag, T. (2013). Challenges in addressing plagiarism in education. *PLoS Medicine, 10*(12), e1001574.

Broeckelman-Post, M. A. (2008). Faculty and student classroom influences on academic dishonesty. *IEEE Transactions on Education, 51*, 206–211. https://doi.org/10.1109/TE.2007.910428

Carrell, S. E., Malmstrom, F. V., & West, J. E. (2008). Peer effects in academic cheating. *Journal of Human Resources, 43*, 173–207. https://doi.org/10.1353/jhr.2008.0013
Eisenberg, J. (2004). To cheat or not to cheat: Effects of moral perspective and situational variables on students' attitudes. *Journal of Moral Education, 33*, 163–178. https://doi.org/10.1080/0305724042000215276
Evans, E. D., & Craig, D. (1990). Adolescent cognitions for cheating as a function of grade level and achievement status. *Journal of Adolescent Research, 5*, 325–345. https://doi.org/10.1177/074355489053005
Festinger, L. (1954). A theory of social comparison processes. *Human Relations, 7*, 117–140. https://doi.org/10.1177/001872675400700202
Galloway, M. K. (2012). Cheating in advantaged high schools: Prevalence, justifications, and possibilities for change. *Ethics & Behavior, 22*, 378–399. https://doi.org/10.1080/10508422.2012.679143
Hartshorne, H., & May, M. A. (1928). *Studies in deceit*. Norwood Press.
Heckler, N. C., Forde, D. R., & Bryan, C. H. (2013). Using writing assignment designs to mitigate plagiarism. *Teaching Sociology, 41*, 94–105. https://doi.org/10.1177/0092055X12461471
Koul, R., Clariana, R. B., Jitgarun, K., & Songsriwittaya, A. (2009). The influence of achievement goal orientation on plagiarism. *Learning and Individual Differences, 19*, 506–512. https://doi.org/10.1016/j.lindif.2009.05.005
Latova, N. V., & Latov, I. V. (2008). Cheating in the educational process. *Russian Education and Society, 50*, 17–43. https://doi.org/10.2753/RES1060-9393500702
Luo, Y., Ma, T., & Deng, Y. (2023). School climate and adolescents' prosocial behavior: The mediating role of perceived social support and resilience. *Frontiers in Psychology, 14*, 1095566. https://doi.org/10.3389/fpsyg.2023.1095566
Magnus, J. R., Poltervich, V. M., Danilov, D. L., & Savvateev, A. V. (2002). Tolerance of cheating: An analysis across countries. *The Journal of Economic Education, 33*, 125–135. https://doi.org/10.1080/00220480209596462
Malesky, A., Grist, C., Poovey, K., & Dennis, N. (2022). The effects of peer influence, honor codes, and personality traits on cheating behavior in a university setting. *Ethics & Behavior, 32*(1), 12–21. https://doi.org/10.1080/10508422.2020.1869006
McCabe, D. L. (2005). It takes a village: Academic dishonesty and educational opportunity. *Liberal Education, 91*(3), 26–31. https://files.eric.ed.gov/fulltext/EJ720381.pdf
McCabe, D. L., Butterfield, K. D., & Trevino, L. K. (2006). Academic dishonesty in graduate business programs: Prevalence, causes, and proposed action. *Academy of Management Learning & Education, 5*(3), 294–305. https://doi.org/10.5465/AMLE.2006.22697018
McCabe, D. L., & Katz, D. (2009). Curbing cheating. *Education Digest, 75*(1), 16–19.
Murdock, T. B., Miller, A. D., & Kohlhardt, J. (2004). Effects of classroom context variables on high school students' judgments of the acceptability and likelihood of cheating. *Journal of Educational Psychology, 96*, 765–777. https://doi.org/10.1037/0022-0663.96.4.765
Murdock, T. B., Beauchamp, A. S., & Hinton, A. M. (2008). Predictors of cheating and cheating attributions: Does classroom context influence cheating and blame for cheating? *European Journal of Psychology of Education, 23*, 477–492. https://doi.org/10.1007/BF03172754
Nora, W. L. Y., & Zhang, K. C. (2010). Motives of cheating among secondary students: The role of self-efficacy and peer influence. *Asia Pacific Educational Review, 11*, 573–584. https://doi.org/10.1007/s12564-010-9104-2
Plata, S., De Guzman, M. A., & Quesada, A. (2023). Emerging research and policy themes on academic integrity in the age of chat GPT and generative AI. *Asian Journal of University Education, 19*(4), 743–758. https://doi.org/10.24191/ajue.v19i4.24697
Schraw, G., Olafson, L., Kuch, F., Lehman, T., Lehman, S., & McCrudden, M. (2007). Interest and academic cheating. In E. M. Anderson & T. B. Murdock (Eds.), *Psychology of academic cheating* (pp. 59–85). Elsevier.
Socscistatistics. (n.d.). *T-Test Calculator for 2 Independent Means*. Social Science Statistics. https://www.socscistatistics.com/tests/studentttest/default2.aspx

Teodorescu, D., & Andrei, R. (2009). Faculty and peer influences on academic integrity: College cheating in Romania. *Higher Education, 57*, 267–282. https://doi.org/10.1007/sl0734-008-9143-3

Tsai, C. (2012). Peer effects on academic cheating among high school students in Taiwan. *Asia Pacific Educational Review, 13*, 147–155. https://doi.org/10.1007/s12564-011-9179-4

Tsvetkova, M., & Macy, M. W. (2014). The social contagion of generosity. *PLoS One, 9*(2), 1–9. https://doi.org/10.1371/journal.pone.0087275

Walker, R. E., Wiemeler, G. E., Procyk, M. R., & Knake, W. P. (1966). The contagion of cheating. *Psychology in the Schools, 3*, 359–360. https://doi.org/10.1002/1520-6807(196610)3:4<359::AID-PITS2310030412>3.0.CO;2-W

Whitley, B. E. (1998). Factors associated with cheating among college students: A review. *Research in Higher Education, 39*, 235–274. https://doi.org/10.1023/A:1018863909149

Whitley, B. E., Nelson, A. B., & Jones, C. J. (1999). Gender differences in cheating attitudes and classroom cheating behavior: A meta-analysis. *Sex Roles, 41*, 657–680. https://doi.org/10.1023/A:1018863909149

Woo, J., Kang, S. W., Kim, H. K., & Park, J. (2018). Contagion of cheating behaviors in online social networks. *IEEE Access, 6*, 29098–29108. https://doi.org/10.1109/ACCESS.2018.2834

Zanetti, C., & Butera, F. (2023). The chronology of collective cheating: A qualitative study of collective dishonesty in academic contexts. *Current Psychology, 42*(32), 27983–27997. https://doi.org/10.1007/s12144-022-03885-3

Zhao, L., Mao, H., Compton, B. J., Peng, J., Fu, G., Fang, F., Heyman, G. D., & Lee, K. (2022). Academic dishonesty and its relations to peer cheating and culture: A meta-analysis of the perceived peer cheating effect. *Educational Research Review, 36*, 100455. https://doi.org/10.1016/j.edurev.2022.100455

Chapter 8
Including Everyone in Academic Integrity: Policies and Teaching Practice to Engage all Students and Staff

Mary Davis

Abstract This chapter, based on my keynote presentation at the ECEIA Conference in Derby, is focused on including everyone in academic integrity. The aim is to raise awareness of the practical ways that inclusion can be built into every aspect of academic integrity policy and practice. The chapter demonstrates that including everyone can be achieved through reflective dialogue, as experienced in the presentation with opportunities for all delegates to discuss key elements of their academic integrity journeys. It also indicates that using inclusive frameworks, notably Universal Design for Learning (UDL), in academic integrity policies, procedures and teaching resources, can make a significant difference to student experience and engagement with academic integrity. Current concerns with student use of AI are examined from an inclusive and ethical perspective, drawing attention to the student voice, providing appropriate access to assistive technology and addressing the digital divide by avoiding disadvantages to any student group. By highlighting inclusion in academic integrity as a critical and achievable goal for everyone in Higher Education, the chapter provides key takeaways and recommendations through a call to action for educators to discuss, prioritise and embed inclusion in every area of their practice.

Keywords Inclusion · Academic integrity · Policies · Teaching · UDL · Artificial intelligence

Introduction

This chapter is focused on including everyone in academic integrity by reflecting on our own academic integrity experiences, and on promoting inclusion in academic integrity policies and teaching practices. The aim of the chapter is to raise

M. Davis (✉)
Arts Humanities and Social Sciences, Oxford Brookes University, Oxford, UK
e-mail: marydavis@brookes.ac.uk

M. Davis, C. Baptista (eds.), *Ethics and Integrity in Education (Practice)*, Ethics and Integrity in Educational Contexts 10,
https://doi.org/10.1007/978-3-031-95686-7_8

awareness of the need for everyone to take a role in promoting academic integrity and to encourage and support greater inclusion to engage all students in academic integrity. I argue throughout that inclusion needs to be prioritised in all academic integrity work through policies, procedures and teaching. Educators need to recognise that academic integrity is an area of knowledge and skills that needs to be taught to all students and reinforced throughout their courses of study (Davis, 2012). I also argue that inclusive approaches are achievable through holistic institutional responses and use of inclusive frameworks, notably Universal Design for Learning, in every area of academic work.

The need for institutions to adopt clear, visible positions on academic integrity through policies and practice is nothing new; already 2 decades ago, Pecorari (2001) highlighted the importance of institutions making academic conduct policies easily accessible and ensuring students' attention is drawn to them at a learning stage, rather than only after a breach. Since then, influential work by Bretag et al. (2011) has led to institutions adopting their guiding principles for exemplary academic integrity policy of access, approach, responsibility, detail and support. Beyond policies, a visible commitment to the practice of academic integrity by institutions, educators and students has been gradually increasing, for example through efforts to build cultures of academic integrity within communities of learning (Reedy et al., 2021), and to inculcate the fundamental values of honesty, trust, fairness, respect, responsibility and courage (ICAI, 2021). However, in the current Higher Education context of major disruption caused firstly by the Covid pandemic and then by the surge of student use of Artificial Intelligence and associated academic conduct concerns affecting stakeholders at every level, it has become much more obvious that everyone needs to play an active role in academic integrity practice.

The chapter will start by outlining my arguments as to why including everyone in academic integrity is crucial to teaching and learning. I will then discuss how I demonstrated including everyone by creating opportunities for reflective dialogue for every member of the audience in the keynote presentation. In the following section, I will focus on embedding inclusion in academic integrity policies and procedures, and then on ensuring teaching practice is inclusive. I will then move to consider current concerns related to inclusion in AI. Finally, I will summarise the main issues related to inclusion in academic integrity and present a call to action with recommendations for all educators to improve inclusion in academic integrity.

The Case for Including Everyone

I argue that it is crucially important to involve everyone in academic integrity, as it is central to student experience, student attainment, professional development, staff experience and institutional reputations, and beyond education has a long-term impact on professions and society. We need to move beyond a classification of academic integrity work as "third space labor" (Ahuna et al., 2023) reserved for a small, dedicated and often under-valued team. In fact, all individuals in Higher

Education, even those who may consider academic integrity to be "a side issue" or outside of their role, need to be actively involved, as declared in the Academic Integrity Charter: "Everyone is responsible as part of a "whole community" approach" (QAA, 2020, p.1). This message is evident in the tagline for the work of ENAI (2024a) "We support higher education institutions to work together in the field of academic integrity". Furthermore, the partner organisation, ICAI (2021) affirms that the academic integrity community needs to "educate all members of the community about academic integrity… (to create) a culture of academic integrity".

Everyone in academia needs to take an active role in inclusion in academic integrity, but especially academic staff as key role models and educators. However, it has been found that academic staff vary significantly in their support of students' development of academic integrity from the "integrity ambassador" who promotes and fully embeds academic integrity into their teaching, to the "detached professor" who completely abdicates their role and responsibility for teaching academic integrity (Peters et al., 2019). Evidently, if all academic staff could be persuaded, prepared and trained to take on the active role of academic integrity ambassador, this would facilitate a holistic approach to teaching and supporting students' development of skills, promoting institutional services, raising awareness, and acting as a key role model to cultivate a community of practice in academic integrity.

As an academic integrity community, we need to reflect on the drivers that have helped us to get where we are, so we can then share, promote and inspire others to join us, so that we advance further. These drivers are the academic integrity networks such as ENAI, the many working groups, the kinds of outreach global events (for example, the International Day of Action for Academic Integrity), the establishment of awards, regular publications in formal journals as well as through blogs and newsletters, all enabling those in academic integrity to take a greater role in Higher Education and to have their voices heard. This movement to recognise the importance of academic integrity has grown significantly in recent years. Although even some colleagues working in academic integrity say that this work is "not their real job", for the most part this narrative is changing and the recognition of the value of academic integrity work is growing. We have been campaigning to #makeitsomeonesjob to advance the professionalisation and prioritisation of academic integrity, the recognition for academic integrity work, and support for setting up academic integrity offices (for example, see the recent work of McKenzie, 2024). Furthermore, the need for every institution to adopt a visible position on academic integrity through policies, procedures and pedagogy has become much more established and widespread, for example as shown through the extensive country-wide and regional surveys conducted by Glendinning (2013, 2016).

Events in recent years have also contributed to a massive escalation of interest in academic integrity across a much broader spectrum of Higher Education colleagues. The remote teaching and assessment forced by Covid lockdowns led to a surge of academic misconduct cases which could not be ignored by faculty (Stoesz et al., 2023). One of the first immediate causes for widespread concern was the increase in the proliferation of contract cheating services offering 24/7 instant exam help or replacements for institutional tutors they claimed were not giving them adequate support remotely. This included huge growth in students colluding or sharing assignments through the use of file sharing sites that provide exam answers (Lancaster & Cotarlan, 2021). At the same time, it was noted that students were suffering from increased stress, anxiety and loneliness resulting from disengagement (Elmer et al., 2020), with students at a distance feeling that they do not belong, and consequently that in a remote setting, it matters less if they cheat. This disruption to students' usual way of learning seems to have contributed to poor decision-making which could include cheating (Stoesz et al., 2023). Their study reported that over half of the students surveyed considered that the imposition of remote instruction led to more academic misconduct among fellow students. All these factors have meant that institutions had to become more active in considering academic integrity more holistically, adapt policies, provide more education for academic integrity, in line with the approach of "pedagogies of integrity" that are integrated in every student's curriculum (McNeill, 2022).

While the years adapting to the disruption of Covid and remote teaching increased the focus on academic integrity, a much more powerful global concern about academic integrity was fueled by the sudden and massive development of Artificial Intelligence (hereafter AI) following the release of ChatGPT3 in November 2022. As argued by Eaton (2023), in this post-plagiarism era, academic staff feel compelled to engage with the ethical questions that AI presents. Thus, a wide and exponentially growing range of educators began to take substantial interest in the academic integrity issues involved in the use of AI, attending events as mass audiences at hitherto unprecedented levels for academic integrity, and furthermore began foregrounding academic integrity in new policies on AI and academic practice. This has dramatically changed perceptions of the importance and reach of academic integrity, so that there is now debate and awareness among a much broader and more varied global audience of educators (Johnston et al., 2024).

Therefore, these events have set in motion a sea change for educators to be more concerned about academic integrity: finally, there is widespread acknowledgement that academic integrity matters to everyone and it has become clearer and easier to provide the evidence. So, we can capitalise on this moment to reinforce the value of academic integrity. But at the same time, academic integrity needs to be accessible to all, otherwise we are missing an opportunity to speak to the value of including everyone in academic integrity where no one is excluded from the conversation or from learning good practice. Thus, I now turn to how I demonstrated a means of including everyone in the keynote presentation.

Reflection on Including Everyone in Practice

The ECEIA Conference in Derby was an occasion to reflect on how interest has grown in academic integrity, as well as the awareness of its relevance and importance to all those working in Higher Education. As a means of demonstrating 'including everyone' in practice, I began the keynote by sharing the key milestones of my academic integrity journey over the last two decades and encouraged delegates to reflect on and share their own experiences, motivations and sense of their roles in academic integrity in one-minute dialogue iterations to answer five questions. Using two of Brookfield's (1995) lenses for critical reflection, namely the autobiographical and the sharing of peer experiences, I believe this kind of active exchange provides an important opportunity for personal and collegiate evaluation of pedagogy and individual contributions to academic integrity, in a powerful, large group, community-building innovation.

I began this process of reflection by asking the question "When and why did you start to take an interest in academic integrity, and what was your first impact?" I wanted to demonstrate that in order to be in that room, at that conference, we had all had a starting point in academic integrity and we had all achieved impacts on academic integrity in our respective contexts and locations. I shared that my starting point had been in 2005 when my students asked me for more feedback on their use of sources at a draft stage, that I had tried out using Turnitin formatively which led to an internally funded project for 3 years and to steering the institutional Turnitin policy towards the development of good practice, rather than punitive measures. I gave the audience 1 min to turn and talk to the person nearest to them and to ask each other the same questions. I observed delegates across the room enthusiastically sharing their experiences and expressing delight in the unusual opportunity for everyone to interact in a keynote, which set the scene and continued for the discussion of each of the subsequent questions.

I followed with two questions about external impacts, to focus on how academic integrity professionals take opportunities to disseminate and bring insights beyond their institutions, again to share experiences, some with challenges. Firstly, to answer the question "What do you remember about your first external presentation on academic integrity?", I shared the hostile reaction I had received from the audience for my first external presentation on the formative use of Turnitin who labelled me as "subversive" for providing my students access to Turnitin before assessment; I considered that including some of the barriers and challenges in my journey would encourage others to share their own and how they dealt with them. Secondly, I asked "What do you remember about your first external publication on academic integrity?" where I recounted a positive collaboration with my mentor, Jude Carroll, and support from Tracey Bretag, the International Journal for Educational Integrity (IJEI) editor at the time. I shared this as I wanted the audience to take a moment to reflect on the colleagues that have supported, mentored and inspired them in the academic integrity community, aware that many audience members would remember and feel indebted to Tracey Bretag.

I then asked about career development, as this is so important to the progress of academic integrity as a field, with the question: "Which of your studies of academic integrity has developed your career the most?" I explained that this was straightforward for me to answer as I had dedicated my 7 years of PhD research to examining how students develop competence in use of sources (citation, paraphrasing, establishing their voice), and how this led to publications, recognised expertise and new roles in academic integrity. I asked one final question to encourage a return to present circumstances and current activities among delegates: "Where are you now? What is your current passion in academic integrity?" and expanded on my commitment to inclusive, educational approaches to academic integrity with the application of Universal Design for Learning (UDL) and contributions to the academic integrity agenda through events, organisations, quality assurance bodies, networks and personal connections. In preparing for this talk, I thought about several other questions, and could have made the entire keynote a reflective experience with the audience but saved this for a future occasion. Instead, I then moved to focus on my own work on inclusive approaches to academic integrity in policies and teaching, as I outline below.

Inclusion in Policies and Procedures

My current work is strongly focused on promoting greater inclusion in academic integrity procedures, policies and teaching. Inclusion has become a priority in Higher Education, and institutions are required to demonstrate how they meet standards of inclusive curricula (Equalities Act, 2010 in Gov.UK, 2013), provide fully accessible resources (Gov.UK, 2021) and act on attainment gaps for specific student groups (Office for Students, 2021). Nevertheless, inclusion in academic integrity is an area that has been overlooked until quite recently, despite continuous evidence that certain groups of students are over-represented in academic conduct investigations (Pecorari, 2016) and need more guidance to address the ongoing award gaps (Awan, 2020). Opportunities to learn about good academic practice have been considered unequal (Wingate, 2015) and some groups such as international postgraduate students have reported that they are left to figure it out for themselves as they are "supposed to know" (Davis, 2012). In these situations, there is an obvious need for greater inclusion in academic integrity; as the ICAI Fundamental Values declare, institutions and educators must provide learning opportunities to everyone in the academic community (ICAI, 2021).

My recent research into inclusion issues (Davis, 2022) revealed that students going through academic conduct investigations may feel disadvantaged for various reasons such as their neurodiversity, coming from an international domicile or having a different prior experience of education. My student participants reported significant anxiety about judgement and a lack of support with learning about academic integrity. I will demonstrate the way I addressed these inclusion issues through making a strong case to change the academic conduct procedure for students with first minor breaches in their first year of study, so that they are provided with an educational route, instead of investigation and punishment.

Problems with academic integrity policies in terms of inclusion clearly need to be addressed, such as location, format and involvement of students in the policy. Many policies are still located in difficult locations that require clicking through many links to find (Stoesz et al., 2019); it needs to be recognised that if we want students to access the policy, it should not be hidden. The format also needs to be accessible, so that it is easy for all students to follow, as presented by Reedy et al. (2021). Their community of practice model, co-created with students, established an innovative and engaging academic conduct process with consistent images and clear messages using flow charts. In my institution, as part of my drive to make academic integrity inclusive, I mapped institutional academic integrity documents to the Universal Design for Learning (UDL) principles for comprehension (CAST, 2018). This led to improvements in consistency and clarity across all documents (see Davis, 2022, 2023), through addressing the four main checkpoints for comprehension in the UDL framework: activating prior knowledge, highlighting relationships, guiding information processing and maximising transfer (CAST, 2018). These changes have meant the documents now incorporate features to assist comprehension for all students, such as standard numbering, key points in bold and consistent terminology. Other examples of positive inclusive practice in academic integrity policies have involved giving students more agency by involving them directly in the creation and revision of academic integrity policy (Khan et al., 2020) and as champions and ambassadors to promote academic integrity to other students in events and training (Gottwald & Milka, 2023).

In considering inclusion in academic integrity policy, it is important to examine the key components of any policy to ensure it meets students' needs. The seminal work by Bretag et al. (2011) set out the five pillars of exemplary academic integrity policy as approach, responsibility, detail, support and evaluation. From an inclusion lens, it is particularly important to recognise "responsibility" as an essential component of policies. Many subsequent studies of policies have used the benchmark standards of Bretag et al.'s work to assess other policies, for example: Anohina-Naumeca et al. (2018) about policies in Latvia and Lithuania; Möller (2022) about New Zealand; Stoesz et al. (2019) based on policies in Canada. Significantly, the work of Eaton et al. (2023) is further extending the exemplary policy principles from Bretag et al. (2011) to raise more awareness of inclusion. Applying these principles to academic integrity policies can help to encourage a global shift in increasing inclusion and recognising student needs.

Inclusion in Teaching

In addition to improving inclusion in policies, I argue that it is equally important to address inclusion in teaching academic integrity, by ensuring all students have learning opportunities in "accessible, relevant and engaging" formats (Thomas & May, 2010). I adopted these three words as my mantra for improving inclusion in academic integrity; therefore, in my work I ensure the format is accessible for all students, the task is clearly connected to their study, and the activity is interesting and interactive for them. Evidence suggests that sometimes staff are unsure how to engage large groups of students in academic integrity, especially if they come from Widening Participation backgrounds as mature students (Fudge et al., 2022). Therefore, I believe it is very important to share and disseminate ways to address inclusion and make sure every student can understand and engage in academic integrity.

To foster greater inclusion in teaching academic integrity at this conference, I presented resources from a QAA-funded collaborative enhancement project that I led, involving four UK HEIs called "The Improvement of student learning by linking accessibility/inclusion and academic integrity" (QAA, 2023). In the presentation, I highlighted the ways to build inclusion into teaching by going through the stages of this collaborative project. The project teaching resources were carefully designed by the project team comprising academic integrity experts working with inclusion experts, as well as Student Union officers and students, so that all these perspectives were included. The resources created are as accessible as possible, firstly checked by inclusion experts, then rechecked using accessibility tools and tested in the institutions with various formats available. The resources include a PARTNERS template (Prior experiences; Accessibility; Reinforcement; Timing; Navigation; Engagement; Revision; Specificity) to assist staff in developing their own resources and checking them against inclusion criteria, and interactive teaching resources for students in the form of games and discussions. The resources are available as open access in order to assist educators in other institutions to improve inclusive practice in academic integrity (QAA, 2023) and are shared on the ENAI educational materials list (ENAI, 2024b). The ENAI list of educational resources is itself an inclusive venture in sharing open access teaching materials on academic integrity in 22 languages other than English, for a wide range of teaching purposes and audiences, including all levels of study and for staff training (ENAI, 2024b). Furthermore, the ENAI working group on educational resources is engaged in promoting and extending the scope of these resources to ensure optimal experience and access to all.

Inclusion and AI

To consider current concerns about inclusion in academic integrity, it is obviously important to look at practices with AI. These concerns are not entirely new as AI has actually been influencing student practice and academic integrity for some time. For example, Rogerson and McCarthy (2017) drew attention to the significance of students using online paraphrasing tools because of the concerns that this can stop students developing their writing skills and learning about their subjects. Other previous studies highlighted concerns with translation tools or grammar tools, especially for institutions to have policies on appropriate use of machine translation, given that they recruit heavily from student populations that do not have a high level of proficiency in English language who may use such tools (Groves & Mundt, 2021). All of these tools are now considered to be under the umbrella of AI tools, which presents some significant inclusion issues. As pointed out by McDermott (2023), if AI is banned or restricted in assignment instructions, students with different learning needs who use AI as an assistive technology will be significantly and disproportionately disadvantaged. New or revised policies must take this inclusion issue into account.

Furthermore, it is the subject of considerable debate, as well as future research, whether the use of free and openly available AI tools could be an opportunity for greater inclusion or whether these tools still disadvantage certain groups due to issues such as paywalls or accessibility. The importance of improving inclusion in academic integrity in relation to AI remains a critical goal for everyone. As highlighted by Rana and Verhoeven (2023), the AI digital divide can exacerbate the already existing barriers to technology; it is essential to ensure accessibility and skills for all students. Important questions relate to whether institutions should provide access to AI, and whether AI democratises by facilitating open access to all or continues to favour privileged populations. Some initiatives have begun to promote the positive, ethical use of GenAI in Higher Education, such as UNESCO (2023) with guidance for a human-centred, positive, equitable use of AI worldwide. Similarly, Illingsworth (2023) argues for the importance of equal access for all students, such as by institutions partnering with community organisations to provide students with access to technology.

Student agency is paramount in the current very new AI climate to inform practice and policy and to ensure all student voices are listened to. The recent UK-based student survey by Johnston et al. (2024) reported that most student participants held the view that universities should have policies on when to use AI, prioritise equal access for all students, and teach use of AI as they recognised its importance for their future careers. The survey found that students with a higher level of confidence were less likely to use AI, and less likely to support other students using AI. It also found that students are happy to support the use of AI assistive technology for students with disabilities, but not for any other student group such as second language speakers of English, as they considered these students may gain an unfair advantage. Ongoing research that foregrounds the student voice in AI is evidently needed to examine these issues of inclusion.

My recent work (Davis, 2024) outlines a positive approach to addressing the need to teach students ethical decision making with AI in an inclusive format. Having gathered an evidence base of what students were actually doing with AI as disclosed through the institutional declaration form system, I categorised these declarations into appropriate, at risk and inappropriate practice. I then used the abovementioned four checkpoints of comprehension in UDL (CAST, 2018) to create an inclusive self-study course through the following steps. In order to connect with background knowledge, I started by explaining the need for guidance on AI and how it is prevalent in HE. To ensure that the course illustrated the connections between concepts, I created a traffic lights table with the three categories of appropriate, at risk and inappropriate practice under the relevant columns. To guide information processing, I used the colours consistently and common symbols in learner settings of tick, question mark and cross for appropriate, at risk and inappropriate. Finally, to provide an opportunity to apply learning, I developed an exercise for students to test their ability to identify appropriate, at risk and inappropriate uses of

AI. The course has been evaluated as fully accessible and meeting inclusive goals (Davis, 2024).

Conclusion

This chapter has discussed the concept of including everyone in academic integrity. It began by reflecting on the need for everyone to take responsibility for academic integrity and demonstrating one way to involve others through a peer and personal reflection on how colleagues interact with academic integrity. It then proceeded to discuss and reflect on the ways to create greater inclusion in academic integrity policies and procedures, and examined inclusive, holistic approaches to teaching academic integrity. Finally, it drew attention to the current necessity for inclusive practice in approaches to AI, providing an example of a course based on UDL to teach ethical decision making with AI.

The chapter has thus provided an overview of the importance of academic integrity for everyone involved in Higher Education and presented ways to engage all students and staff in academic integrity positively and inclusively. By prioritising inclusion in policies, procedures, and teaching in all areas of academic integrity, including those related to AI, institutions will not only meet inclusivity goals, but also improve the experience of academic integrity for all. The call for action below sets out in brief the recommended actions for everyone to play a role in inclusive academic integrity.

Call to Action: Recommendations for Including Everyone in Academic Integrity

1. Make academic integrity 'relevant, accessible and engaging' for all students and staff;
2. Ensure that academic integrity is talked about in relation to all areas of study and curriculum;
3. Change the narrative by correcting those who say academic integrity is a "side issue" or "not a real job" and explaining to them that it is central to the success of Higher Education;
4. Ensure that academic integrity policies are fully inclusive by incorporating Universal Design for Learning;
5. Ensure that teaching academic integrity is built into every programme and accessible for every student;
6. Draw attention to the need to take an inclusive approach to academic integrity in connection with AI.

References

Anohina-Naumeca, A., Tauginienė, L., & Odineca, T. (2018). Academic integrity policies of Baltic state-financed universities in online public spaces. *International Journal for Educational Integrity, 14*(8), https://doi.org/10.1007/s40979-018-0031-z.

Ahuna, K., Frankovitch, L., & Murphy, G. (2023). Claiming space for honest work: Academic integrity as third space labor. *Workplace: A Journal for Academic Labor, 34*, 75–87. https://ices.library.ubc.ca/index.php/workplace/article/view/186840.

Awan, R. (2020, December 21). The awarding gap at the Open University. The Open University: OpenLearn. https://www.open.edu/openlearn/education-development/the-awarding-gap-the-open-university.

Bretag, T., Mahmud, S., Wallace, M., Walker, R., James, C., Green, M., East, J., McGowan, U., & Partridge, L. (2011). Core elements of exemplary academic integrity policy in Australian higher education. *International Journal for Educational Integrity, 7*(2), 3–12. https://doi.org/10.21913/IJEI.v7i2.759

Brookfield, S. (1995). *Becoming a critically reflective teacher*. Jossey-Bass.

CAST. (2018). Universal Design for Learning Guidelines version 2.2. http://udlguidelines.cast.org.

Davis, M. (2012). International postgraduate students' experiences of plagiarism education in the UK: Student, tutor and expert perspectives. *International Journal for Educational Integrity, 8*(2), 21–33. https://ojs.unisa.edu.au/index.php/IJEI/article/view/807/601.

Davis, M. (2022). Examining and improving inclusive practice in institutional academic integrity policies, procedures, teaching and support. *International Journal of Educational Integrity, 18*(14). https://doi.org/10.1007/s40979-022-00108-x

Davis, M. (2023). Inclusion within a holistic approach to academic integrity: Improving policy, pedagogy, and wider practice for all students. In S. E. Eaton (Ed.), *Handbook of academic integrity* (pp. 1–19). Springer. https://doi.org/10.1007/978-981-287-079-7_127-1

Davis, M. (2024). Supporting inclusion in academic integrity in the age of GenAI. In J. Lawrence, S. Beckingham, P. Hartley, & S. Powell (Eds.), *Using GenAI effectively in HE* (pp. 21–30). Routledge. https://doi.org/10.4324/9781003482918

Eaton, S. E. (2023). Postplagiarism: Transdisciplinary ethics and integrity in the age of artificial intelligence and neurotechnology. *International Journal for Educational Integrity, 19*(23). https://doi.org/10.1007/s40979-023-00144-1

Eaton, S. E., Vogt, L., Seeland, J., & Stoesz, B. M. (2023, June 1-2). *Academic integrity policy analysis of Alberta and Manitoba colleges* [Conference presentation]. Winnipeg, MB, Canada: Canadian Symposium on Academic Integrity (CSAI), University of Manitoba. https://hdl.handle.net/1880/116575.

Elmer, T., Mepham, K., & Stadtfeld, C. (2020). Students under lockdown: Comparisons of students' social networks and mental health before and during the COVID-19 crisis in Switzerland. *PLoS One, 15*(7), 1–22. https://doi.org/10.1371/journal.pone.0236337

ENAI (2024a, April 4). European Network for Academic Integrity: Home page. https://www.academicintegrity.eu/wp/.

ENAI (2024b, April 4). European Network for Academic Integrity: Educational materials list. https://academicintegrity.eu/materials/.

Fudge, A., Ulpen, T., Bilic, S., Picard, M., & Carter, C. (2022). Does an educative approach work? A reflective case study of how two Australian higher education enabling programs support students and staff uphold a responsible culture of academic integrity. *International Journal for Educational Integrity, 18*(5). https://doi.org/10.1007/s40979-021-00099-1

Glendinning, I. (2013, October). Comparison of policies for academic integrity in higher education across the European Union. Impact of Policies for Plagiarism in Higher Education Across Europe (IPPHEAE) Project Consortium. http://plagiarism.cz/ippheae/files/D2-3-00%20EU%20IPPHEAE%20CU%20Survey%20EU-wide%20report.pdf.

Glendinning, I. (2016). European perspectives of academic integrity. In T. Bretag (Ed.), *Handbook of academic integrity* (pp. 1–16). Springer Science + Business Media. https://doi.org/10.1007/978-981-287-079-7_3-2
Gottwald, C. & Milka, A. (2023, March 10). Students as leaders*: Developing an academic integrity ambassador programme* [Conference presentation]. International Centre for Academic Integrity Conference, Indiana, USA.
Gov.UK. (2021, April 4). CDDO accessibility regulations campaign information for education. Central Digital and Data Office. https://www.gov.uk/government/publications/online-accessibility-regulations-campaign-supporter-pack/gds-accessibility-regulations-campaign-information-for-education.
Gov.UK. (2013). Equality Act 2010. https://www.gov.uk/guidance/equality-act-2010-guidance.
Groves, M., & Mundt, K. (2021). A ghostwriter in the machine? Attitudes of academic staff towards machine translation use in internationalised higher education. *Journal of English for Academic Purposes, 50*, 1475–1585. https://doi.org/10.1016/j.jeap.2021.100957
Illingsworth, S. (2023, May 18). If AI is to become a key tool in education, access has to be equal. *The Conversation*. https://theconversation.com/if-ai-is-to-become-a-key-tool-in-education-access-has-to-be-equal-204156#comment_2913535.
ICAI (2021). T*he fundamental values of academic integrity* (3rd ed.). Available under a creative commons attribution-NonCommercialShareAlike 4.0 international license. https://academicintegrity.org/images/pdfs/20019_ICAI-Fundamental-Values_R12.pdf.
Johnston, H., Wells, R. F., Shanks, E. M., Boey, T., & Parsons, B. N. (2024). Student perspectives on the use of generative artificial intelligence technologies in higher education. *International Journal for Educational Integrity, 20*(2). https://doi.org/10.1007/s40979-024-00149-4
Khan, Z. R., Mumtaz, S., & Rakhman, S. S. (2020). Mentoring to affect student perceptions of academic integrity. In D. Velliaris (Ed.), *Academic language and learning support services in higher education* (pp. 98–118). IGI Global. https://doi.org/10.4018/978-1-7998-2878-5.ch004
Lancaster, T., & Cotarlan, C. (2021). Contract cheating by STEM students through a file sharing website: A Covid-19 pandemic perspective. *International Journal for Educational Integrity, 17*(3). https://doi.org/10.1007/s40979-021-00070-0
McDermott, B. (2023, April 5). *QQI artificial intelligence week: Blurred lines Rethinking AI tools through assistive technology* [Video]. YouTube. https://www.youtube.com/watch?v=ZN7kxgxCkAU.
McKenzie, A. (2024, March 9). *#Makeitsomeonesjob: Building an academic integrity office* [Conference presentation]. International Center for Academic Integrity Conference, Calgary, Canada.
McNeill, L. (2022). Changing "Hearts" and minds: Pedagogical and institutional practices to Foster academic integrity. In S. E. Eaton & J. Christensen Hughes (Eds.), *Academic integrity in Canada. Ethics and integrity in educational contexts* (pp. 487–503). Springer. https://doi.org/10.1007/978-3-030-83255-1_25
Möller, A. (2022). An analysis of university academic integrity policies in New Zealand. *Journal of Further and Higher Education, 47*(3), 338–350. https://doi.org/10.1080/0309877X.2022.2130195
Office for Students. (2021, November 9). Degree attainment*:* Black, Asian and minority ethnic students. https://www.officeforstudents.org.uk/advice-and-guidance/promoting-equal-opportunities/effective-practice/black-asian-and-minority-ethnic-students/.
Pecorari, D. (2016). Plagiarism, international students and the second-language writer. In T. Bretag (Ed.), *Handbook of academic integrity* (pp. 537–550). Springer.
Pecorari, D. (2001). Plagiarism and international students: How the English-speaking university responds. In D. Belcher & A. Hirvela (Eds.), *Linking literacies: Perspectives on L2 Reading-writing connections* (pp. 229–245). University of Michigan Press.
Peters, M., Boies, T., & Morin, S. (2019). Teaching academic integrity in Quebec universities: Roles professors adopt. *Frontiers in Education, 4*, 99. https://doi.org/10.3389/feduc.2019.00099

QAA. (2020, October 21). *Academic integrity charter for UK higher education*. https://www.qaa.ac.uk/membership/membership-areas-of-work/academic-integrity/charter.

QAA. (2023). The improvement of student learning by linking inclusion/accessibility and academic integrity. Collaborative Enhancement Project—Oxford Brookes University, University of Southampton, Bloomsbury Institute and University of Westminster. https://www.qaa.ac.uk//en/membership/collaborative-enhancement-projects/academic-integrity/the-improvement-of-student-learning-by-linking-inclusion-accessibility-and-academic-integrity.

Rana, V., & Verhoeven, B. (2023). How to bridge the artificial intelligence divide. LSE Business Review. https://blogs.lse.ac.uk/businessreview/2023/08/04/how-to-bridge-the-artificial-intelligence-divide/.

Reedy, A. K., Wurm, P. A. S., Janssen, A., & Lockley, A. (2021). A community of practice approach to enhancing academic integrity policy translation: A case study. *International Journal for Educational Integrity, 17*(12). https://doi.org/10.1007/s40979-021-00080-y

Rogerson, A. M., & McCarthy, G. (2017). Using internet based paraphrasing tools: Original work, patchwriting or facilitated plagiarism? *International Journal for Educational Integrity, 13*(2), https://doi.org/10.1007/s40979-016-0013-y.

Stoesz, B. M., Eaton, S. E., Miron, J., & Thacker, E. J. (2019). Academic integrity and contract cheating policy analysis of colleges in Ontario, Canada. *International Journal for Educational Integrity, 15*(4). https://doi.org/10.1007/s40979-019-0042-4

Stoesz, B. M., Quesnel, M., & De Jaeger, A. E. (2023). Student perceptions of academic misconduct amongst their peers during the rapid transition to remote instruction. *International Journal for Educational Integrity, 19*(14). https://doi.org/10.1007/s40979-023-00136-1

Thomas, L., & May, H. (2010, September 1). Inclusive learning and teaching in higher education. HE Academy. https://www.advance-he.ac.uk/knowledge-hub/inclusive-learning-and-teaching-higher-education.

UNESCO. (2023). *Guidance for generative AI in education and research*. UNESCDOC Digital Library https://doi.org/10.54675/EWZM9535.

Wingate, U. (2015). *Academic literacy and student diversity: The case for inclusive practice*. Multilingual Matters.

Chapter 9
Faculty Perspectives on Academic Dishonesty: Evolving Beliefs and Opportunities for Learning at Two Canadian Research Universities

Lydia Scholle-Cotton and Ruth A. Childs

Abstract Understanding faculty members' experiences with student academic dishonesty and how these experiences relate to their current attitudes, beliefs, and behaviours is important for Higher Education, where institutional reputation depends on institutions effectively discouraging dishonest behaviour. This study used semi-structured interviews with nine faculty members at two universities in Ontario, Canada, to investigate faculty members' experiences with attitudes and beliefs toward, and actions in response to students' academic dishonesty. The interviews included several survey questions. Due to the richness of explanations provided by the faculty, a thematic analysis was used. A second analysis applied the Theory of Planned Behaviour to gain additional insights into faculty members' behaviours and actions when dealing with cases of academic dishonesty. The study finds that faculty attitudes, beliefs, and actions evolve based on experience and familiarity with existing policies. Furthermore, most faculty members prioritised working with students to help them understand academic dishonesty and how to avoid it. Faculty members varied in their use of formal channels to deal with academic dishonesty cases. Those who had previously served in administrative roles and had more exposure to institutional policies were more likely to feel comfortable using formal systems than those who had not.

Keywords Academic dishonesty · Cheating · Theory of planned behaviour · Higher education · Discussion making · Faculty-student interaction

L. Scholle-Cotton (✉)
Queen's University, Kingston, ON, Canada
e-mail: lydia.cotton@queensu.ca

R. A. Childs
University of Toronto, Toronto, ON, Canada
e-mail: ruth.childs@utoronto.ca

M. Davis, C. Baptista (eds.), *Ethics and Integrity in Education (Practice)*, Ethics and Integrity in Educational Contexts 10,
https://doi.org/10.1007/978-3-031-95686-7_9

Introduction

Universities are committed to promoting academic integrity among students. Lapses in academic integrity are often associated with students' behaviour in relation to cheating or plagiarism (Macfarlane et al., 2014). While most universities continue to improve and promote processes around academic integrity, policies and procedures are often created based on cases that have been formally reported for disciplinary action. This focus on formally reported cases overlooks unreported cases. While many faculty members follow established rules about reporting academic integrity cases, the perceived challenge of proving instances of academic dishonesty by students may discourage some faculty members from reporting cases (Coalter et al., 2007). As a result, our understanding of faculty members' experiences with students cheating remains incomplete, particularly concerning the academic dishonesty cases they routinely encounter but do not report.

McCabe et al. (2010), in their summary of research on students' academic dishonesty, note that students who observe other students cheating also observe faculty members seemingly ignoring such behaviours. Possible reasons that faculty members ignore cheating include that they do not recognize it, they do not know how to respond, they do not believe that responding is worthwhile, and they are dealing with additional issues around the introduction of Generative AI. Multiple studies have also identified differences between the beliefs of students and beliefs of faculty regarding behaviours that should be deemed as cheating, their severity and prevalence (Anderman & Won, 2019; Naghdipour & Emeagwali, 2013). Wajda-Johnston et al. (2001) found that, even at the graduate level, students were more likely than faculty members to recognize that their behaviour constituted academic dishonesty.

Understanding differences in perceptions becomes even more important when considering faculty members' multiple roles within the university. Faculty members often have responsibilities as educators, researchers, mentors, and administrators. Moreover, as educators, faculty members play a vital role in nurturing essential writing and critical thinking skills among students. However, this responsibility has become increasingly complex due to varying levels of student preparedness (Côté et al., 2020; Grayson et al., 2019). These challenges are further exacerbated by the far-reaching impacts of the COVID-19 pandemic, which has reduced the academic preparedness of some students (Aurini & Davies, 2021).

A gap exists in our understanding of faculty members' experiences with academic dishonesty. Better understanding faculty members' experiences with student academic dishonesty and how these experiences relate to their current attitudes, beliefs, and responses to new cases is important for Higher Education, where institutional reputation depends heavily on the ability of an institution to discourage academically dishonest behaviours.

In this study, nine faculty members from two prominent Canadian research universities were interviewed about their beliefs, attitudes, experiences, and actions regarding student academic dishonesty. The thematic analysis showed widespread consistency in faculty members' beliefs regarding the importance of preventing cheating and a tendency, when faced with an instance of cheating, to treat it as a learning opportunity for students in the hope of preventing the student from engaging in similar behaviour in the future. Where faculty members differed in their use of formal channels when addressing cases of academic dishonesty, those who had held more administrative roles within their institutions, were more likely to rely on formal systems. A second analysis helped to explain faculty members' actions when dealing with cases by linking faculty members' behaviour to the Theory of Planned Behaviour (Ajzen, 1991).

This study advances our understanding of how faculty members are primarily working with graduate students, highlighting their experiences, attitudes, and behaviour.

Method

This study used semi-structured interviews that included several survey questions to investigate how faculty members respond to students' academic dishonesty. The research focused on two key aspects: (1) the beliefs of faculty members regarding the frequency and challenges of academic dishonesty and (2) how their past experiences influence their beliefs and actions. The data were analysed in two ways. In the first analysis, a thematic analysis was performed using Braun and Clarke's (2006) six-step approach. In the second analysis, the themes identified in the first analysis were matched with the Theory of Planned Behaviour (Ajzen, 1991).

Participants

A purposive sampling method (i.e. non-random selection) was used, and faculty were selected from two universities with Faculties of Education with graduate programs. Both universities were ranked within the top 5 in the Medical Doctoral category of Canadian universities with PhD programs and medical schools (Maclean's, 2020). Faculty selection within these institutions was based on the stage of their academic career (years at the university as indicated on their publicly-available curriculum vitae) and gender to ensure the sample included variation on both dimensions. Ten faculty members were emailed in the Winter of 2022 asking if they were willing to participate in the study. Of the 10, nine responded and agreed to participate. Participation was voluntary, with both emailed and verbal consent provided. Participants were not compensated.

Nine faculty members from two universities in Ontario, Canada, participated in this study. All participants have taught classes in the field of education. Seven participants were tenured or tenure track; two were lecturers or contract faculty. While not all the participants were currently teaching classes, all faculty members have taught graduate courses at some point within the last 4 years, and three faculty taught both undergraduate and graduate courses. This homogeneity of faculty from the same discipline (Education) and having experience teaching graduate students allowed the opportunity to identify meaningful themes (Braun & Clarke, 2006). However, there was some heterogeneity in the wide range of teaching experience, ranging from less than 4 years to more than 20 years. Most faculty indicated that they had additional responsibilities such as serving on committees and several had served in academic leadership positions, such as Department Chair or Associate Dean. Four participants identified as male and five as female; although gender was recorded, there were no substantial differences in thematic analysis by gender. It is important to point out that this study was conducted in the Winter of 2022, before the release of easily accessible Generative AI applications such as OpenAI ChatGPT in November 2022.

Survey Questions

The semi-structured interview (see Appendix A) included survey questions about faculty members' experiences in addressing academic dishonesty. These questions were developed, following a comprehensive review of existing literature (Anderman & Won, 2019; Coalter et al., 2007; Naghdipour & Emeagwali, 2013), to include a variety of dishonest behaviours by students, such as plagiarism, copying exam answers, reproducing another student's assignment, and using unauthorised materials. The survey questions were refined and clarified through cognitive interviews before the study.

Data Collection

The interviews were conducted by the first author via a "live Zoom meeting" and lasted between 30 and 50 min. Before the recording began, the researcher reviewed the general information sheet, and the participant provided informed consent. After reviewing both documents, the researcher asked the participant if it was okay to begin recording. At the beginning of the recording, the researcher again confirmed that it was okay to proceed with the interview and confirmed with the participant that they had reviewed the general information sheet, provided informed consent, and was willing to proceed. Each survey question posted by the interviewer in the Zoom chat feature and read out loud for the participant.

Participants were invited to qualify and/or expand on their response to each of the survey questions and all participants did expand on most of their responses to the survey questions. In addition, some participants discussed their experiences with students' academic dishonesty more broadly. Transcription via the Zoom platform was enabled. Following each interview, the transcript was reviewed, compared with the recording, and corrected when necessary. Identifying information was removed. After each interview, the first author wrote field notes that included observations about the interview, such as any interruptions, sound issues, and the emotional tone of the interview.

Before conducting the study, the research protocol was approved by the University of Toronto's Social Sciences, Humanities, and Education Research Ethics Board, Protocol # 42254.

Data Analysis

Braun and Clarke's (2006) six-step approach to conducting a thematic analysis helped guide the thematic analysis:

1. **Get comfortable with the data**. The first author reviewed each transcript at least three times to ensure the transcript was accurate and included the appropriate level of data. Any identifying information was removed or altered to ensure participant anonymity.
2. **Generate the initial codes**. This step involved identifying interesting characteristics of each interview.
3. **Search for initial themes within the data**. This step involved considering additional themes that may not have been present in the initial research question but are emerging from the data.
4. **Review and validate the initial themes**. Having identified potential themes across respondents, the nine transcripts were reviewed again to identify all evidence (and potential counterpoints) to the proposed themes, determining which are supported by robust evidence from the data.
5. **Organise and name the identified themes**. This step involved organising and restructuring the data to ensure it illustrated a well-organised story about the data collected and clearly spoke to the themes.
6. **Produce a final report of the data**. This step involved selecting representative and illustrative examples from the dataset to relate to the analysis findings and research questions.

Aligning the themes from the first analysis with the Theory of Planned Behaviour (TPB) as a framework encompassing attitude, subjective norms, and perceived behaviour provided a systematic approach to investigating faculty members' decision-making processes when addressing academic dishonesty. Most notably, Coren (2012) has used the TPB to gauge faculty members' willingness to address students' cheating. Coren's study was based on survey responses from faculty members and demonstrated that the TPB model can offer insights into the overall climate and attitudes within the faculty community. In particular, Coren found that faculty attitudes predicted whether a faculty member met with a student. While the TPB has most often been used in quantitative studies (e.g. Ababneh et al., 2022; Coren, 2012; Stone et al., 2009), several qualitative studies have also demonstrated its usefulness (King & Dennis, 2006; McTaggart-Cowan et al., 2021; Renzi & Klobas, 2008).

Results

The initial thematic analysis yielded three themes across the participants: (1) faculty learning through experience, (2) unconfirmed cases, and (3) corrective discipline and learning opportunities for students. As will be described in the following sections, faculty members adapted their behaviours in response to students engaged in academic dishonesty. This adaptation underscores the dynamic nature of faculty members' attitudes, beliefs, and disciplinary measures, which continually evolve based on prior experiences and familiarity with existing policies.

Faculty Learning Through Experience

One of the interview questions asked participants to reflect on their first experience with academic dishonesty. All participants indicated they had experiences with academic dishonesty while teaching, but how they dealt with their first experience varied, as presented in Table 9.1.

A follow-up question asked participants what actions they would take "today" if they were convinced a student had engaged in academic dishonesty. The responses are presented in Table 9.2. Comparing Tables 9.1 and 9.2, we notice that two faculty members say they are more likely than before to report the incident to their chair or dean, and three faculty members are more likely to follow the policy. It should be noted that not all participants answered all questions; therefore, the total counts may not be the same.

These results may suggest that faculty members, as they gain more experience with academic dishonesty or take on academic leadership roles, tend to become more acquainted with the existing policy and are more inclined to adhere to it. A study by MacLeod and Eaton (2020) supports this finding: They found that most faculty are familiar with existing policy and were made aware after incidents with

Table 9.1 Responses to their first experience with academic dishonesty

	Participant									
	1	2	3	4	5	6	7	8	9	Total
Allowed the student to redo exam/assignment				1	1				1	3
Lowered the student's grade				1	1		1			3
Gave a failing grade for the course	1		1	1	1	1				5
Reported the incident to my Chair/Dean			1	1			1	1		4
Followed the policy as set by the institution	1		1	1						3
Did nothing about the incident										0
Gave a warning							1			1
Other		1								1
Total	2	1	3	5	3	1	3	1	1	20

Table 9.2 How they would respond to academic dishonesty now

	Participant									
	1	2	3	4	5	6	7	8	9	Total
Allow the student to redo exam/assignment									1	1
Lower the student's grade										0
Give a failing grade for the course			1							1
Report the incident to my Chair/Dean	1	1	1		1	1		1		6
Follow the policy as set by the institution	1	1	1		1		1	1		6
Do nothing about the incident										0
Give a warning										0
Other				1						1
Total	2	2	3	1	2	1	1	2	1	15

students or emails from administration reminding them to reference the policy in their course syllabi.

Participant 1, for example, indicated that they have served in other offices within the university, which affected their expectations about students engaging in academic dishonesty, noting that "if you see it in one place you probably see it in others." Participant 7 had recently been involved in modifying the existing policy and indicated that this experience has increased their knowledge of the policy, saying, "Now I would follow the set policy because we have recently revised the policy, and I am aware of the policy more so now." Similarly, Participant 8 had experience in an academic leadership role and said, "I would [follow the set policy], I know the policy now a little bit more [now] and so I would still report them [the student to the chair] but I would also then follow the set policy."

Faculty members also described how their approaches to academic dishonesty had changed over time, including changes in how they teach. For example, Participant 5 said, "I was very careful in the kinds of assignments I gave to students, so I would be able to get a sense if they were building their final research paper, which would be the place where cheating might take place." Participant 9 mentioned incorporating text-matching software in courses. Participant 2 felt that requiring students to prepare portfolios made it difficult for students to cheat. Participant 7 also described how they have changed their evaluation methods:

> I actually have changed many of my assignments to be applied work with communities and organizations in relevant areas of the graduate training, and that takes plagiarism right out of the way because actually the test of the course and the knowledge is actually applying it in real-world settings and collaborating with others to apply that knowledge. (Participant 7)

Some participants worried that the COVID-19 pandemic had made cheating more tempting for students. For example, Participant 1 had begun surveying students about their workload. They were also concerned about the challenges of learning during the pandemic, especially for international students, and wanted to make sure to provide opportunities to meet with a student who was suspected of academic dishonesty:

> I would likely also want to make sure that there's an opportunity to meet with the student. Then, pending what information emerges if an individual presents extenuating circumstances… I understand they're humans and [have] very challenging circumstances particularly right now. I would consider allowing them to redo [the assignment]. (Participant 1)

In summary, the faculty members in this study positioned themselves as perpetual learners, drawing insights from past experiences and integrating new software tools into their teaching. This approach equips them with updated knowledge of existing policies, enhancing their adeptness in guiding disciplinary actions when necessary.

Unconfirmed Cases

The second theme was the faculty are not always able to identify when a student is cheating. Participants reported seeing strange or weird work that they believe is not in the student's own voice, because it was either written by others or copied from other sources. Or they see work that does not directly match the assignment, potentially reflecting the reuse of past work from other courses. In such cases, most faculty will not pursue formal disciplinary actions but will rather use the situation as a learning opportunity. This also suggests that academic dishonesty is likely more widespread than what is suggested by the numbers of reported cases.

Participant 5 indicated that they use the text-matching software Turnitin but would like to better understand how it can be helpful in their classes, saying, "I think there's a big gray zone in terms of … how much, what percentage of replication that we're willing to accept". Participant 2 described, when submitted work did not match the assignment, worrying that graduate students might be recycling work from their undergraduate degree. Participant 3 reported being suspicious that students might be using external writers, but finding it difficult to confirm. Indeed, Participant 9 was recently reminded of the ease with which students can engage in cheating when they were doing research online and had a pop-up window appear, asking "Do you need a paper written on that?"

In summary, the participants in this study reflected on instances when they suspected academic dishonesty but did not have the concrete evidence required for formal disciplinary action. The difficulty of proving academic dishonesty may help to explain why faculty often choose to try to educate students about academic dishonesty instead of pursuing corrective discipline.

Corrective Discipline and Learning Opportunities for Students

Faculty members were also likely to use incidents of academic dishonesty as opportunities for student learning. The faculty in this study were teaching graduate students in Faculties of Education and reported that the most prevalent form of academic dishonesty they encountered was related to academic writing. Several faculty members perceived a low likelihood of encountering academic dishonesty in their classes, deeming academic dishonesty unlikely or irrelevant.

Faculty members were asked how likely they thought specific types of academic dishonesty had been in their classes in the preceding 4 years. The types listed included plagiarism, copying answers on exams, employing external writers, submitting another student's assignment, using unauthorised materials during exams, fabricating references, manipulating data, or attempting to gain special consideration through favours. For plagiarism, in particular, three faculty members indicated it was "Highly Likely", one gave a "Neutral," response, three chose "Unlikely," and two chose "Highly Unlikely".

During the interview, faculty members were asked their perceptions of why students cheat. Most faculty members believed that academic dishonesty is not a result of too much leniency and unclear rules or class plans. In particular, seven of the nine faculty members "Strongly Disagreed" that the university did not make the rules clear, "Strongly Disagreed" that the university's rules were not strict enough, and "Strongly Disagreed" that the class plan was not clear.

In the interviews, faculty indicated that they are responding to academic dishonesty as a learning opportunity for students. For example, Participant 4, who had taught for more than 20 years, suggested that the action taken by the faculty should depend on how egregious the student's academic dishonesty was. They explained what they do in their classes: "When it sounds to me like the student's voice in what I'm reading isn't theirs, I point that out in the feedback that I give them and explain to them how important it is to recognize the voices of those whom they're drawing on".

Participant 2 recalled a very memorable experience working as a Teaching Assistant with a senior faculty member who was reluctant to acknowledge that a student had plagiarised one of the assignments. This experience led them to believe that it is important for students to understand what they did wrong and understand how to correct it. They described their current view:

> I actually really feel strongly that students don't learn anything from us saying, "Oh, you're caught, and the student vanishes." People learn a whole lot when you tell them that they have to redo work because... this is work. (Participant 2)

Participant 4 also believed cheating often occurs because of students misunderstanding what is allowed and described how they try to teach the students about academic dishonesty through special sessions on how to reference. Participant 3 identified the use of text-matching software as a learning opportunity for students, suggesting that, "When they turn in their assignment, it goes through the tool and they see the scores provided by the tool and then they take remedial action".

Participant 9 expanded on how students can learn what is acceptable and what is not. They suggested a training course or certificate course, explaining, "Through that kind of scenario-based module or training and capacity building, I think that it will be much clearer to students what is and is not plagiarism, because I don't think it's always intentional".

Participant 5 noted that they believed that misunderstandings of plagiarism were especially prevalent among students without undergraduate training in North America, saying, "The real reason they plagiarise is they don't fully understand what plagiarism means, and they don't understand plagiarism, particularly in the North American context if they're not educated in or have done their undergrad in North America." This participant suggested that when learning a new language people often repeat back or copy what others say or write and this may be at the core of the issue.

While many of the participants believed that students do not understand what constitutes academic dishonesty, they also believed that the policies at their institution and the statements on course syllabuses clearly defined academic dishonesty. This could reflect a disconnection between policies and the communication of the policies, which might be addressed through training on the topic as some participants suggested.

Several participants described feeling upset by the ways they had seen cases of academic dishonesty addressed. For example, Participant 4 described an incident from early in their career as a faculty member, when they felt another faculty member had gotten caught up in how to implement and interpret the rules:

> It was stupid, it was just so clear that this was the wrong thing to do. Because the student hadn't been dishonest, but they got caught up in the way we implement the rules or interpret the rules. This student's life was destroyed by this.

Participant 6 remembered an international exchange student not completing their doctorate because of the consequences of an academic dishonesty case:

> In the end, he didn't get his doctorate and he moved to another part of the country. But I think he found a future for himself. His international experience was quite helpful, so it didn't destroy his whole career. But it was taken very seriously at the time. So, that was my one very direct and painful experience of academic dishonesty.

Participant 5 also described a difficult situation that they still thought about:

> One student, I failed him, and … he had to delay his graduation by a whole semester, from his Master's [degree], and it was really traumatic for him, I think because his whole family was supposed to accompany him to his graduation. So, he had to take a whole other class in order to graduate on time. And yet he also felt ashamed, because he shouldn't have plagiarised the whole paper. And so, he didn't harass me or anything, he just accepted the consequences, and it was sad. I think about it.

In summary, faculty members often perceive instances of academic dishonesty as learning opportunities for the student involved, attributing cheating to potential misunderstandings rather than intentional misconduct. When they do see students receive corrective discipline, the faculty members may be upset if they perceive the effects of the discipline to be disproportionate to the students' actions.

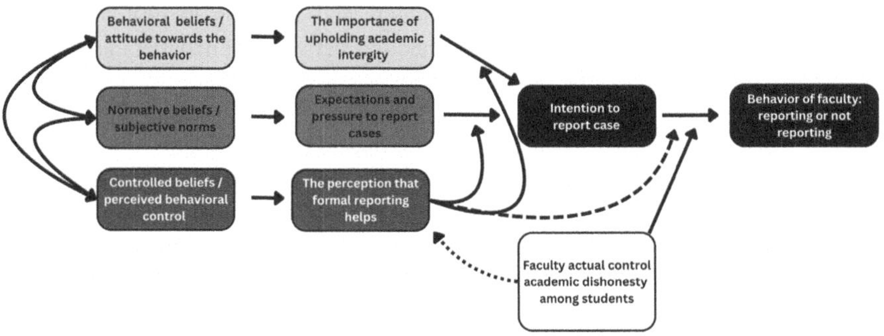

Fig. 9.1 The theory of planned behaviour (adapted from Ajzen, 1991)

Alignment with the Theory of Planned Behaviour

The TPB framework provides insights into faculty members' navigation through complex situations, such as when deciding how to address a suspected case of academic dishonesty. The framework describes the relationships among personal beliefs, social influences, and faculty members' perceived sense of control, helping to explain their subsequent actions and behaviours. In the subsequent sections, an in-depth exploration of each component within the TPB model will follow, elucidating its relevance and significance within the specific context of our study. Figure 9.1 illustrates how the TPB is interpreted in this study. The framework remains unchanged from the original, except for adding descriptions and text related to this study. This section of the paper provides a detailed discussion of how each of the belief inputs (Behavioural beliefs/Attitudes Towards Behaviour, Normative Beliefs/ Subject Norms, and Controlled Beliefs/Perceived Behavioural Control) in the framework are connected to the analysis and ultimately how these beliefs influence the decision-making of faculty members when dealing with cases of academic dishonesty.

Behavioural Beliefs/Attitude Towards the Behaviour

Connecting the thematic analysis to the TPB framework, the initial linkage pertains to the concept of "attitude". As used by Ajzen (1991), this term encapsulates a comprehensive evaluation of behaviour, indicating whether an individual is inclined or disinclined to engage in a specific behaviour. As shown in Fig. 9.1, behavioural beliefs have a direct impact on a faculty member's attitudes. The interview began by asking participants if upholding the academic integrity of their institution was an important part of their role as faculty members who teach; the majority of the participants answered "Strongly Agree" or "Agree". The TPB framework suggests that faculty members with a positive regard for academic integrity would inherently recognize the imperative of reporting incidents and adhering to protocols when addressing academic dishonesty. This connection between their beliefs and anticipated behaviour highlights the profound influence of their attitudes on their subsequent actions.

Yet, in contrast to this, the impact of past experiences and outcomes emerged through the theme of "learning through experience", and, in many cases, these experiences highlight underlying compassion that faculty may have towards students they encounter when dealing with academic dishonesty. As faculty members embrace their responsibility to preserve academic integrity, they may become more or less inclined to support institutional policies. This correlation emphasises the pivotal role that individual faculty members' views play in the collective effort to uphold the integrity of the educational institution and how, as they grow in their roles, their past experiences and outcomes become a meaningful part of their roles as faculty members.

Normative Beliefs and Subject Norms

Exploring the concept of subjective norms, as introduced by Ajzen (1991), reveals the influence of social pressures on the adoption or avoidance of certain behaviours. These pressures stem from an individual's alignment with their normative beliefs or the prevailing influence exerted by their social networks or groups. While our research did not explicitly inquire about faculty members' exposure to social pressures, a consistent trend emerged: Many faculty members emphasised instances of unconfirmed cases, suggesting the potential for seeking advice from colleagues or peers and the perceived pressure from the administration to adhere to existing policies. For example, Participant 1 talked about the lack of opportunities to discuss possible cases with colleagues during the pandemic, noting that, "there's less of an opportunity to come across someone and say "hey, oh, can I pick your brain?"

The faculty members' focus on fixing problems instead of just disciplining students shows how faculty members' beliefs, their jobs, and what the academic community expects all come together. When faculty members choose to fix things rather than strictly follow rules, it is important to think about how they see the existing policies.

It is important to remember that faculty members may feel pressures from society that affect their choices about reporting cheating. These pressures can come from colleagues, the administration, or even the rules and policies that are already there. By investigating how familiar faculty members are with the policies and how much they trust them, we may be able to reduce some of these societal pressures. This could help faculty feel more comfortable and confident about what to do in these situations.

Controlled Beliefs and Perceived Behaviour Control

In the TBP, controlled beliefs are how individuals see their control over a situation and their ability to act as they want (Ajzen, 1991). In this study, when we asked faculty members if they would take formal action now when encountering academic dishonesty, 3 chose "Strongly Disagree," 3 chose "Neutral," 1 chose "Agree," and 2 chose "Strongly Agree." This might suggest that these cases of academic dishonesty matter to faculty members, showing a possible difficulty in using existing policies to manage student behaviour. However, the variation in responses could also mean that, as faculty spend more time at the institution, deal with academic dishonesty more, or take on senior roles, they become more familiar with the policy and are more likely to follow it.

Adding more complexity, sometimes participants felt that they or their colleagues erred in trying to apply the academic integrity policy, harming students in the process. The diversity of responses and the strength of faculty members' feelings about cases underscore the intricate interplay among faculty members' perceptions, institutional policies, and the ramifications of academic dishonesty. While some instances might be amenable to educational interventions, others present hurdles that faculty members persistently contend with, underscoring the necessity for a holistic and compassionate approach to academic integrity.

Furthermore, there is a call for a more inclusive and effective strategy for addressing academic dishonesty that empowers faculty to feel a sense of control. This approach aims to foster a harmonious balance between maintaining academic standards and supporting faculty members in promoting integrity.

Discussion

Universities struggle to maintain academic integrity as they face challenges posed by declining literacy skills, varying student preparedness, the ongoing impacts of the COVID-19 pandemic, and now student access to Generative AI. Preventing and responding to cases of academic dishonesty requires that institutions establish rules and procedures and faculty and staff consistently apply these procedures. A key concern is that faculty, who are integral to upholding academic integrity, may apply different standards or break from institutional processes when responding to student misconduct cases.

There exists a gap in our understanding of faculty experiences with academic misconduct and how they perceive and respond to cases among their students. This study helps close this gap by investigating faculty perceptions and responses when confronted with academic dishonesty. The faculty participants in this study provided valuable insights into their experiences and beliefs regarding academic dishonesty. While they varied in teaching experience and roles within their institutions, common themes emerged, shedding light on their attitudes, normative beliefs, and control beliefs.

Firstly, faculty members recognized the importance of upholding academic integrity within their institutions. This finding is consistent across faculty in both teaching and administrative roles. This alignment of perspectives underscores the common view within institutions that preventing and addressing student misconduct is a priority. There is wide consensus that institutions should work to safeguard academic integrity and that faculty play an important role in maintaining a trustworthy academic environment. However, if faculty members are not bringing cases forward, this is not necessarily because they do not see maintaining academic integrity as a priority.

Secondly, faculty members highlighted that they often faced cases of academic dishonesty that they could not prove and they were reluctant to pursue formal disciplinary actions in such situations. This reluctance to report such cases is often reinforced by conversations with peers and a general perception that reporting is time-consuming without any benefits to the faculty member or the student. This is true even when the rules require such cases to be reported. This observation highlights the complex interplay of faculty members' normative beliefs, professional roles, and societal influences when responding to academic misconduct.

Lastly, some faculty members expressed a reluctance to report students because that meant turning over control of the outcome to the administration and potentially subjecting the student to a harsher punishment than the faculty member believed was appropriate. This is consistent with the idea of controlled beliefs, as defined in the Theory of Planned Behaviour, where the faculty members' perceptions of control over situations influenced their readiness to initiate formal disciplinary actions. In some cases, this led to faculty members not following procedures and attempting to deal with misconduct on their own. This complexity emphasises the need for a comprehensive and empathetic approach to academic integrity.

Together, these insights suggest that due to their perceptions of the effectiveness and fairness of the procedures, faculty members often break from them. However, we also found evidence that more experienced faculty members and those who have participated in the administration were more likely to follow the rules and procedures. This suggests that the procedures may not be the problem but rather the faculty's lack of information or knowledge of existing procedures and policies. This highlights the potential benefits of faculty training on what is expected of them in a wide array of situations they may face.

Faculty members' attitudes, beliefs, and disciplinary actions regarding academic dishonesty are dynamic and continually evolving based on prior experiences and exposure to existing policies. Administrative offices may be able to support faculty by providing spaces for informal discussions on academic dishonesty and by offering consistent teaching methodologies for addressing this issue. It may be important to ensure that more experienced faculty members participate in these spaces. Further research is needed to expand these findings to other academic fields and explore the graduate student-faculty relationship in the context of academic dishonesty.

Ultimately, fostering a culture of academic integrity is a collective effort that requires ongoing dialogue and collaboration between faculty and students.

Our study took place in the months before the public release of ChatGPT. The emergence of Generative AI applications and their use among students adds to the pressures that universities are currently facing to establish effective and efficient procedures for faculty to deal with cases of academic misconduct.

Recommendations

The following recommendations to establish faculty roles in academic integrity emerge from this study:

1. Involve faculty members in the development of academic integrity procedures.
2. Ensure that faculty members are fully informed about these procedures.
3. Foster the belief among faculty that adherence to these procedures is essential.
4. Promote collaborative efforts among administration, faculty, and students in the development of academic integrity policies and guidelines.

We highlight that these recommendations are increasingly important in light of emerging technologies, such as Generative AI, which further highlight the need for thoughtful and inclusive policy development.

Limitations of This Study

The Theory of Planned Behaviour (TPB) is typically employed in quantitative studies, where surveys are often tailored to measure each of the framework's components. In this study, in contrast, the TPB framework was applied to the results of the thematic analysis of the interview responses. This use of the TPB framework, post-analysis, introduced an additional lens to interpret the data, facilitating a deeper understanding of the faculty members' behaviours and their driving factors.

Furthermore, this research focused on faculty members educating graduate students in the Faculties of Education. An examination of the existing literature reveals a lack of research specifically addressing academic dishonesty at the graduate level, particularly within the context of Canada. However, future studies will be needed to investigate academic dishonesty among graduate students in other disciplines and types of intuitions. This study used a qualitative approach with semi-structured interviews. Future research could expand the study to more participants by modifying the existing survey (see appendix); this would support more quantitative analyses.

Appendix

Survey Questions

Background Questions

- Please state your current Faculty position [select one]
- Non-tenure, continuing adjunct, lecture, term adjunct, contract faculty, tenure track [assistant professor, associate professor, professor, other [fill in the blank]]
- Indicate your academic discipline
  - Creative Arts, Humanities, Languages, Physical Sciences, Social Sciences, Law, Business, Engineering, Medicine, other [fill in the blank]
- Department
  - [fill in the blank]
- How many years have you taught a class(es) at the university level as a faculty member?
  - 0, 1-3, 4-5, 6-9, 10-19, 20+
  - If 0, then explain why [interview ends]
- What level class do you currently teach [select all that apply]
  - Undergraduate, graduate, post-grad, other [fill in the blank]
- Please select the gender that best describes you
  - Male, female, non-binary, two-spirited, prefer not to say, other [fill in the blank]
- How many hours do you typically teach per week [select one]
  - 0-4, 5-8, other [fill in range]
- If you have additional university responsibilities, please select all that apply
  - Chair, director, committee(s), research chair, other [fill in the blank]

Survey Questions

- Upholding the academic integrity of the institution is an important part of my role as a faculty member who teaches.
- Strongly Disagree, Disagree, Neutral, Agree, Strongly Agree
- Thinking back to your first experience with a student and academic dishonesty while teaching, please select all that apply.
  - Allowed the student to redo exam/assignment
  - Lowered the student grade
  - Gave a failing grade for the course
  - Reported the incident to my Chair/Dean

  - Followed the set policy as set by the institution
  - Did nothing about the incident
  - Gave a warning
  - Other [fill in the blank]

- Thinking about your first experience of academic dishonesty, to the best of my knowledge the student was [respond to all]
  - Student's role: Undergraduate, graduate, distance student, don't know or can't remember
  - Student's international status: International or domestic student, don't know or can't remember
  - Gender: Male, female, non-binary, two-spirited, prefer not to say, other [fill in the blank]
  - Student's level: First-year, upper-year, mature student, don't know or can't remember, other [fill in the blank]

- Now, if you were convinced a student had engaged in academic dishonesty, select the following action(s) you are most likely to take [select all that apply]
  - Allow the student to redo exam/assignment
  - Lower the student grade
  - Give a failing grade for the course
  - Report the incident to my Chair/Dean
  - Follow the set policy as set by the institution
  - Do nothing about the incident
  - Give a warning
  - Other [fill in the blank]

- In the last four academic years, how likely do you think the following occurred in your class(es)? For each statement below, indicate the likelihood of the action occurring by a student(s) ["not relevant for my courses" or 1-5 with 1 = Extremely unlikely and 5 = Highly likely]
  - Plagiarism on any assignment
  - Copying exam answers
  - Copying another student's assignment and turning it in
  - Using unauthorized materials during a quiz
  - Using unauthorized materials during an exam/final
  - Use of essay farm or external writers
  - Fabricating references or a bibliography
  - Altering data (e.g., adjusting data to obtain a significant result)
  - Attempting to obtain special consideration by offering or receiving favours through, for example, bribery, seduction, corruption
  - Other [fill in the blank]

- What type of delivery modes are you most likely to teach?
  - Before COVID-19 (before September 1, 2019), I taught: In-person only, mix online and in-person, online only
  - During the beginning of COVID-19 (September 1, 2019 – August 31, 2021), I taught: In-person only, mix online and in-person, online only
  - Now, I'm teaching: In-person only, mix online and in-person, online only
- Thinking through your *most recent* experience of a student engaging in academic dishonesty, indicate your agreement, for each statement below, what you believe is the reason the student engaged in the activity ["not relevant for my courses" or 1-5 with 1 = Extremely unlikely and 5 = Highly likely]
  - The university did not have clear policies on cheating
  - The university did not have tough policies when students cheat
  - The syllabus was unclear on what the guidelines were
  - Peer pressure
  - To increase their GPA or score
  - The student is not at the university to learn but simply to get a 'certificate'
  - Cheating is normalized for the student, they have always done it, so they continue to do it
  - Was not able to manage time well
  - Other [fill in the blank]
- Based on your experience dealing with students and academic dishonesty throughout your career, please indicate how in agreement you are with this statement: *My willingness to pursue formal disciplinary action against those who cheat in my courses has increased.*
  - Strongly Disagree, Disagree, Neutral, Agree, Strongly Agree
- In the last year, while teaching, how many distinct students engaging in academic dishonesty have you experienced?
  - [fill in the blank]
- Have you changed your evaluation methods for students to deter any possible academic dishonesty? [select one]
  - Have not changed at all
  - Altered a little
  - Have made major changes
- Have you changed your syllabus for your class(es) to deter any possible academic dishonesty? [select one]
  - Have not changed at all
  - Altered a little
  - Have made major changes

- Based on your experience with academic dishonesty and students would you say you are now [select one]
  - More likely to up-hold the existing policies
  - Haven't changed my view
  - Less likely to up-hold the existing policies
- Based on your experience with academic dishonesty and students, how strongly do you think your institution supports faculty members when dealing with these issues? [select one]
  - Could do a lot more
  - There is some support but there could be more to support faculty
  - There is enough support provided
  - If you think your institution could do more to support faculty, what would you like them to do? [fill in the blank]
- Have you ever pursued official reports (or made an official report to the administration) about a case of academic dishonesty?
  - Yes, no, I never had a strong enough case, prefer not to answer
  - If no, why not? [fill in the blank]
  - If yes, why? [fill in the blank]
  - How many years ago was the first time? [fill in the blank]
  - How many years ago was the most recent time? [fill in the blank]
  - Would you do it again? [fill in the blank]
- Are there any other memorable instances of student academic dishonesty or experience you wish to share?
  - Yes, no, prefer not to answer
  - If yes, please explain [fill in the blank]

References

Ababneh, K. I., Ahmed, K., & Dedousis, E. (2022). Predictors of cheating in online exams among business students during the Covid pandemic: Testing the theory of planned behavior. *The International Journal of Management Education, 20*(3), 100713. https://doi.org/10.1016/j.ijme.2022.100713

Ajzen, I. (1991). The theory of planned behavior. *Organizational Behavior and Human Decision Processes, 50*(2), 179–211. https://doi.org/10.1016/0749-5978(91)90020-T

Anderman, E. M., & Won, S. (2019). Academic cheating in disliked classes. *Ethics & Behavior, 29*(1), 1–22. https://doi.org/10.1080/10508422.2017.1373648

Aurini, J., & Davies, S. (2021). COVID-19 school closures and educational achievement gaps in Canada: Lessons from Ontario summer learning research. *Canadian Review of Sociology, 58*(2), 165–185. https://doi.org/10.1111/cars.12334

Braun, V., & Clarke, V. (2006). Using thematic analysis in psychology. *Qualitative Research in Psychology, 3*(2), 77–101. https://doi.org/10.1191/1478088706qp063oa

Coalter, T., Lim, C. L., & Wanorie, T. (2007). Factors that influence faculty actions: A study on faculty responses to academic dishonesty. *International Journal for the Scholarship of Teaching and Learning, 1*(1), 10.20429/ijsotl.2007.010112.

Coren, A. (2012). The theory of planned behaviour: Will faculty confront students who cheat? *Journal of Academic Ethics, 10*, 171–184. https://doi.org/10.1007/s10805-012-9162-7

Côté, J., Grayson, J. P., Roberts, S., & Chen, L. (2020). Beyond HEQCO's skills assessments: Validation of a measure of students' academic capital and evidence of grade inflation in Ontario's high schools and universities. *Canadian Journal of Educational Administration and Policy, 194*. Article 194. https://journalhosting.ucalgary.ca/index.php/cjeap/article/view/69016

Grayson, J. P., Côté, J., Chen, L., Kenedy, R., & Roberts, S. (2019). *Academic skill deficiencies in four Ontario universities*. York University.

King, T., & Dennis, C. (2006). Unethical consumers: Deshopping behaviour using the qualitative analysis of theory of planned behaviour and accompanied (de)shopping. *Qualitative Market Research: An International Journal, 9*(3), 282–296. https://doi.org/10.1108/13522750610671699

Maclean's. (2020). Canada's best medical doctoral universities: Rankings. (2019, October 3). https://macleans.ca/education/university-rankings-2020-canadas-top-comprehensive-schools/.

Macfarlane, B., Zhang, J., & Pun, A. (2014). Academic integrity: A review of the literature: Studies in higher education. *Studies in Higher Education, 39*(2), 339–358. https://doi.org/10.1080/03075079.2012.709495

MacLeod, P. D., & Eaton, S. E. (2020). The paradox of faculty attitudes toward student violations of academic integrity. *Journal of Academic Ethics, 18*(4), 347–362. https://doi.org/10.1007/s10805-020-09363-4

McCabe, D. L., Trevino, L. K., & Butterfield, K. D. (2010). Cheating in academic institutions: A decade of research. *Ethics & Behavior, 11*(3), 219–232. https://doi.org/10.1207/S15327019EB1103_2

McTaggart-Cowan, H., Bentley, C., Raymakers, A., Metcalfe, R., Hawley, P., & Peacock, S. (2021). Understanding cancer survivors' reasons to medicate with cannabis: A qualitative study based on the theory of planned behavior. *Cancer Medicine, 10*(1), 396–404. https://doi.org/10.1002/cam4.3536

Naghdipour, B., & Emeagwali, O. L. (2013). Students' justifications for academic dishonesty: Call for action. *Procedia: Social and Behavioral Sciences, 83*, 261–265. https://doi.org/10.1016/j.sbspro.2013.06.051

Renzi, S., & Klobas, J. (2008). *Using the theory of planned behavior with qualitative research* (working paper 12). Carlo F. Dondena Centre for Research on Social Dynamics.

Stone, T. H., Jawahar, I. M., & Kisamore, J. L. (2009). Using the theory of planned behavior and cheating justifications to predict academic misconduct. *Career Development International, 14*(3), 221–241. https://doi.org/10.1108/13620430910966415

Wajda-Johnston, V. A., Handal, P. J., Brawer, P. A., & Fabricatore, A. N. (2001). Academic dishonesty at the graduate level. *Ethics & Behavior, 11*(3), 287–305. https://doi.org/10.1207/S15327019EB1103_7

Chapter 10
Enhancing Academic Practice: A Case Study on Postgraduate Business Student and Staff Expectations and Experiences with Academic Integrity and Artificial Intelligence

Debbie De

Abstract Academic integrity is becoming increasingly complex for universities to manage with developments in Artificial Intelligence (AI) software including ChatGPT and online plagiarism checkers. Through a combination of online surveys, focus groups and 1-1 semi-structured interviews, this case study seeks to enhance teaching and learning practices and policies by exploring student and staff experiences and viewpoints on academic integrity and the use of AI in learning and assessment.

The study shows that international student diversity and differing educational backgrounds are key factors which determine student understanding of academic integrity. Whilst use of AI technologies is becoming increasingly difficult to detect, authentic assessments are identified as a means of mitigating plagiarism and thereby maintaining the credibility of Higher Education (HE). As universities are tasked with ensuring that students develop their digital AI literacy, inclusive of ethical use of AI, this study points to the critical role of Learning Development in helping students navigate the challenges within this dynamic HE environment.

Keywords Academic practice · Academic integrity · Artificial intelligence · Learning development · Turnitin

---

D. De (✉)
Aston University, Birmingham, UK
e-mail: d.de@aston.ac.uk

M. Davis, C. Baptista (eds.), *Ethics and Integrity in Education (Practice)*, Ethics and Integrity in Educational Contexts 10,
https://doi.org/10.1007/978-3-031-95686-7_10

Introduction

The global issue of academic offences relating to academic integrity, particularly plagiarism, has increased with the shift towards online learning and examinations (Liles, 2019; Mullholland, 2020; Noorbehbahani et al., 2022; Yazici et al., 2023).

Alongside rapid developments in Artificial Intelligence (AI) generative technologies, including ChatGPT and Google Gemini (formerly Bard), it is becoming increasingly challenging for universities to detect and mitigate against poor academic practice, including plagiarism and collusion (Guo et al., 2023; Krügel et al., 2023). In the UK context, the Quality Assurance Agency (QAA) points to the gravity of this issue (QAA, 2023a) which has the potential to damage the reputation of the UK's world class Higher Education (HE) sector. It is imperative for educators to develop effective strategies, policies, guidelines and practices to develop academic practice, inclusive of academic integrity and digital AI literacy in order to protect academic standards.

Driven by a philosophy to 'empower students to make sense of academic practices' (ALDinHE, 2022), Learning Development (LD) practitioners are third space workers, who work across and between the academic and social spheres (Whitchurch, 2015; Livingston & Ling, 2022). Accordingly, LD practitioners have a unique insight into student experiences, inclusive of the uncertainties and challenges relating to the ethical use of AI in learning and assessment; the increasing number and range of such queries motivated this study. Developing the knowledge and understanding of LD practitioners in this area through research could be vital in shaping learning and teaching practices relating to AI. Moreover, LD activities which encourage discussion and understanding of the ethics of using AI technologies have the potential to improve academic practice across student cohorts, whilst maintaining the integrity of UK HE qualifications within a globally competitive market.

For ethical reasons, this case study uses pseudonyms to refer to research participants and the institution involved as a means of protecting the identity of participants given the potential scope of discussions on academic practice, academic integrity, including academic misconduct. Thus, with a focus on students and staff from the College of Business and Social Sciences at a Midlands based University (MB University henceforth), this case study discusses postgraduate student and staff expectations, experiences and needs surrounding academic integrity. There is a diverse range of programmes within the College of Business and Social Sciences spanning a range of disciplines, this study focuses on students across the range of business courses, inclusive of Human Resources, Business Analytics and Business and Management studies.

The overarching aim of this study is to enhance the teaching and learning experience for students and staff across the institution by exploring the following research questions:

(i) What awareness, experiences and expectations of academic integrity do students have?

(ii) What practices are students adopting to better understand and ensure effective academic practice?
(iii) How are teaching staff trying to increase student awareness and understanding of academic integrity?

Methods

Qualitative approaches permit a deep understanding of people's interpretations and reflections on their surrounding environment (Adler & Adler, 1987). Driven by a qualitative approach, case study methodology can "contribute to our knowledge of individual, group, organisational, social, political and related phenomena" (Yin, 2009, p.4). This resonates with the aim of this study, which seeks to better understand student expectations and staff practices relating to academic practice in the context of the College of Business and Social Sciences at MB University.

Research Design

There are multiple definitions and metaphors associated with research design, and I adopt Creswell's (2009) holistic interpretation whereby research design is considered a locus of philosophy, strategies and specific methods. This compliments case study methodology, which permits the use and triangulation of multiple research methods (Yin, 2009), in this case: online surveys, focus groups and semi-structured interviews.

Online Surveys

Surveys generate data that can describe and/or explain characteristics, behaviours, attitudes or opinions (Tanner, 2018, cited in Williamson & Johanson, 2018). An online survey was chosen for ease of dissemination and completion to facilitate a high response rate; of the 950 postgraduate students and 365 staff across the College of Business and Social Sciences, a total of 104 students and 47 staff completed the online surveys. Appendices 1 and 2 outline the student and staff online survey questions respectively, which were designed to generate data to increase understanding of participant status and background, views, experiences and understanding. Survey data was used to inform the line of questioning in student focus groups and staff semi-structured interviews discussed below.

Focus Groups

Driven by a qualitative approach, focus groups involve interactive discussions centred on specific topics with the intention of generating a range of views (Hennink, 2014) and generate "group meanings associated with a given issue" (Bloor et al., 2001, p. 7). Despite the risk of individual voices dominating any focus group, focus groups have the potential to expose a range of perspectives and provide insights through nuanced group discussions. See appendix 3 for focus group discussion points.

This study includes 6 focus groups with a total of 36 students. One of the focus groups comprised of 6 students who had previously attended academic misconduct meetings; 5 focus groups were conducted with students without experience of academic misconduct.

Semi-Structured Interviews

Qualitative interviews provide researchers with access to the world of the participant by providing an insight into how individuals "make sense of themselves" and "their experiences" (Silverman, 2016, p. 52). Semi-structured interviews were chosen due to their flexibility, which permitted the researcher to adapt questioning as appropriate (Brinkmann & Kvale, 2015). See appendix 4 for the interview schedule; questions were informed by student focus group data and the researcher understanding of the HE and university context.

A total of 6 semi-structured interviews with university staff working with Postgraduate Business students were conducted. The range of roles of staff participants (see Table 10.1 below) was designed to form a holistic view on staff perspectives on the issue, their perceptions of student needs and how these students are supported at programme level.

Ethics

Pseudonyms are used throughout this paper to refer to the institution which this case study is based on, MB University, and in reference to research participants. Informed consent was obtained from participants; participants received both written and

Table 10.1 Roles of semi-structured interview participants

Staff role	Number of participants
Teaching staff	2
Learning developer	1
Postgraduate careers consultant	1
Subject librarian	1
Student union representative	1

verbal explanations of the purpose of the research and had opportunities to ask questions and clarify their understanding of the study, including the potential use of and dissemination of research findings. Participants confirmed their understanding and willingness to participate in this study by signing the consent forms on the understanding that their identity would be anonymised via pseudonyms. The decision to use pseudonyms was taken to encourage participants to share their authentic academic practices, which may differ from university regulations relating to use of AI and academic integrity.

Thematic Analysis

Thematic analysis, whereby themes are identified from data, was used for exploring, interpreting and reporting patterns of meaning across datasets (Braun & Clarke, 2022). Thematic analysis is used to provide data summaries and descriptions of viewpoints expressed by participants which have the potential to identify and convey nuanced and insightful analysis (Braun & Clarke, 2022).

In this study, focus group and interview transcriptions were reviewed multiple times and the following key themes were identified:

(i) International students and diversity in educational background and academic practice
(ii) Pedagogical implications of AI software
(iii) Turnitin, feedback and LD
(iv) Assessment practices

Results and Discussion

Theme 1: International Students, Diversity in Educational Background and Academic Practice

95 of the 104 students who completed the online survey were international students. Student survey data attributes a lack of confidence, differing practices and uncertainty with referencing requirements to diversity in educational backgrounds resulting from diversity in nationalities. Similarly, interview and online survey data shows that staff also strongly consider diversity among international students and in educational backgrounds as key factors which determine student awareness and understanding of academic practice, particularly academic integrity.

These findings attribute a lack of student understanding of academic integrity and plagiarism with diversity in educational backgrounds resonate with existing research into the area. For instance, Pecorari (2010) discusses the Chinese cultural perspective which views imitation as a form of flattery and honour which may be perceived as plagiarism in Western cultures. More generally, research shows that students from collectivist cultures are less likely to acknowledge work produced by individuals in their writing whilst viewing it their duty as students to repeat the works of others (Pecorari, 2010; McMahon, 2018). Furthermore, individuals accustomed to rote learning involving memorisation as a form of knowledge capital are likely to inadvertently adopt reproductive writing strategies (Pecorari, 2010) which are associated with poor academic practice and represent a "deficit" in the Western educational context (McMahon, 2018).

At the time of data collection, all student participants confirmed their confidence and understanding of the principles of academic integrity. This points to the effectiveness of the university's learning and teaching activities, inclusive of LD activities aiming to develop student awareness and understanding of academic practice. However, each student dataset also reveals a lack of confidence in the specific requirements of referencing and point to a need for ongoing academic support in this area.

Research shows that students are likely to implement academic practices which served them well prior to university unless training or instruction is provided (Draper & Fisher, 2020). This is a significant consideration for universities given the potential diversity of educational qualifications and capital across the student body given the range of routes into universities, such as A Levels and BTECs within the home student population alone. Instruction during the transition into university study is therefore imperative, particularly surrounding academic skills and clarifying expectations with referencing and plagiarism (QAA, 2022).

Arguably, it is critical for universities to "get it right" when it comes to supporting postgraduate master's students transitioning to HE due to the tight time frame of the full-time programmes. Particularly as the majority of students on postgraduate programmes are international students with different experiences of academic practice, including academic integrity (Carroll & Ryan, 2005; McMahon, 2018). All university educators have a role to play here; however, with a duty to enable students to "make sense of academic practices" (ALDinHE, 2022), the work of LD teams is critical. In this context, LD teams have the potential to improve academic practice, increase retention and reduce academic offences at an institution-wide level and thereby contribute to the overall student experience.

Theme 2: Pedagogical Implications of AI Software

The majority of student participants use AI technologies including Grammarly and ChatGPT. Interesting student practices with AI were uncovered, for instance, students collectively purchase and share Grammarly premium accounts. Furthermore, student participants pointed to the ubiquity and everyday usage of AI software. One participant explained that ChatGPT was a tool in her everyday life and therefore should be embedded within her formal learning experiences. Another student participant went as far as stating "I'm guilty, I'm in love with these three things: Grammarly, Quillbot and ChatGPT".

It is worth noting that despite the global increase in usage of ChatGPT (Guo et al., 2023; Krügel et al., 2023), at the time of conducting the focus group, not all student participants were aware of the AI software. However, such is the rapidity in development of AI technologies, the more sophisticated ChatGPT-4 which requires a fee-paying account is less detectable than earlier free versions (Dickenson, 2023; Staton, 2023). Student practices of sharing Grammarly accounts disclosed during data collection may extend to ChatGPT-4 at the time of writing this paper a matter of months after data collection.

There is consensus among staff participants that it is the role of educators to teach students how to use AI effectively as a tool for learning, thereby encouraging ethical, effective and transparency in use of AI software in assessments. This finding resonates with Russell Group (2023), which, despite initial criticisms of use of generative AI in HE, advocate the ethical use of AI whilst maintaining academic rigour in teaching and learning. Similarly, the QAA (2023a) advises universities to 'work with' these technologies rather than attempt to ban their usage; advocating clear policies and guidelines to ensure ethical academic practice, enhance student understanding of digital AI literacy and to protect the integrity of university programmes.

Theme 3: Turnitin, Feedback and LD

In terms of detecting cheating, data shows that staff are confident in their ability to interpret Turnitin reports; however, many staff put forward the need for further training in using Turnitin software. Specifically, there is a lack of consistent practice across the university in permitting students to access Turnitin reports prior to assignment deadlines as a means of identifying areas of good and weaker academic practice relating to use of evidence and referencing. Students explained that inconsistent use of Turnitin across the university, particularly surrounding access to Turnitin and Turnitin reports prior to assessment deadlines constitutes a form of unfairness and this is something that the university must address.

All 6 student participants with a history of academic misconduct felt that university communications regarding notification of potential misconduct and invitation to appeals were insensitive and anxiety inducing. As international students, notification of potential misconduct led to feelings of shame and fear of deportation. Awareness of the stress associated with these accusations and a rise in the identification of false accusations of plagiarism is one of the reasons why the majority of UK institutions rejected Turnitin's AI detector tool (Staton, 2023). Other reasons for rejecting the AI detector tool include the hasty and seemingly forced introduction of the tool, preventing universities from learning how to use the software effectively, and Turnitin's admission that the tool was incapable of detecting use of ChatGPT-4, i.e. the paid version of the software (Dickenson, 2023).

Aware of the potential of Turnitin as a tool for academic development (Mphahlele & McKenna, 2019), staff participants voice their desire to make better use of the software. Indicative of a lack of staff resources and large student cohorts, teaching staff participants attribute their busy schedules and the resulting shortage of time to their inability to support students interpret their Turnitin reports. However, data also shows a lack of staff awareness of existing teaching and learning resources which could address this challenge. The LD interviewee discussed existing online resources, including staff and student co-created resources designed to inform students about academic practice, including how to interpret Turnitin reports and develop academic practice accordingly. Furthermore, some teaching staff were also unaware that LD staff can provide taught sessions at module level, including interactive workshops involving Student Writing Coaches to discuss Turnitin reports and feedback more generally. Such activities could be beneficial for teaching staff, in terms of reducing workload whilst improving academic success. Despite a significant university-wide increase in LD teaching at programme level, the ongoing lack of awareness of the remit and potential of LD professionals in universities is a frustration which many LD professionals may be familiar with.

Theme 4: Assessment Practices

The majority of staff participants cited the increase in online exams as the most critical factor contributing to cheating in assessments. Both student and staff participants point to the scheduling of exams and coursework as a stress inducing factor which can push students towards online essay mills or the use of AI software to help with assessment. These findings are in line with Case et al. (2019) longitudinal study on cheating with business students which found an increase in online learning and assessments exacerbated cheating.

Staff data also points to the potential of redesigning assessments, in particular, a move towards authentic assessments and this is indicative of university-wide discourses on assessment. To most staff interviewees, authentic assessment refers to an imitation of tasks required by employers, such as consultancy projects in real-time and viva-type exams. Such assessments would require up to date information and the application and analysis of data in a specific context, which AI software is currently unable to produce, thereby reducing the potential of unethical use of AI in assessment (QAA, 2023b).

Though not explicitly in the context of online learning, Ashford-Rowe et al. (2013) determined critical elements of authentic learning which they used to develop a framework for authentic assessment. In addition to basic features of assessment, such as evidencing the transfer of knowledge, their framework for authentic assessments includes opportunities for collaboration, an assessed "final product" or "performance" and metacognition, to determine the level of critical reflection, including self-evaluation. These are skills which are valued by employers and may imitate professional tasks. At present, AI is not yet equipped to effectively manage the practical nature, requirement for collaboration, communication and reflection associated with authentic assessment as such. Therefore, the concept of authentic assessment and authentic learning could be critical for those working at all levels of education globally. Across HE, an institution's ability to ensure authentic learning and assessment could prove to be integral in maintaining the reputation of the institution and the value of HE qualifications.

Conclusion and Recommendations

Through an exploration of postgraduate business student and university staff perspectives and experiences, this study has highlighted important themes relating to academic practice in the context of rapidly developing AI technologies, namely: educational capital relating to experience and understanding of academic integrity, AI in assessment and authentic assessment.

Diversity in educational capital remains a key consideration for educators, particularly whilst students are transitioning to university. It is imperative for universities to provide clear information regarding the expectations of academic practice, especially regarding academic integrity, plagiarism and referencing requirements during this phase of the student journey. Use of AI represents an emerging concern in education and globally, there are different views on the acceptable use of AI technologies in assessment. In the UK context, the QAA (2023a) encourages universities to provide students and staff with clear guidance on the ethical use of AI technologies. However, with the challenges of detecting generative AI technologies in assessment, authentic assessment has been presented as a means of limiting opportunities for cheating.

This study shows that the role of LD professionals is critical in ensuring students and staff develop their understanding and feel supported with the challenges surrounding referencing, ethical use of AI and digital AI literacy. At an institutional level, this study found that despite the range of LD activities and learning resources available to students and staff, there is a lack of awareness and under-usage of LD resources. This highlights a need for greater promotion and an ongoing dialogue between the LD team at MB University and staff across the institution to ensure that students are better supported, and teaching staff may work more effectively.

More generally, though LD activities vary across and within institutions, within this dynamic sector, LD professionals have a responsibility to participate in institutional and sector-wide discourses on academic practice. In this way, we can help shape future policies and regulations which are accessible to students and staff; directly contributing to an increase in academic confidence and success whilst maintaining academic standards within the sector.

Appendices

Appendix 1: Online Student Survey Questions

1. What is your student status?
    - Home student (UK)
    - European student
    - International student
    - Prefer not to say

2. In which country did you carry out your undergraduate studies?
3. At the start of your postgraduate course, what academic issues/skills were you most concerned with?
4. How would you rate your understanding of assessment criteria at the start of your postgraduate course?
5. How would you rate your understanding of academic integrity at the start of your postgraduate course?
6. How would you rate your understanding of the importance of academic integrity at the start of your postgraduate course?
7. How would you rate your understanding of plagiarism at the start of your postgraduate course?
8. How would you rate your awareness of the consequences and dangers of plagiarism at the start of your postgraduate course?
9. How would you rate your understanding of referencing at the start of your postgraduate course?
10. Which – if any – of the following webinars did you attend?

    - Avoiding academic offences
    - Writing skills – how to paraphrase and use evidence in assignments
    - How to reference
    - Developing academic practice
    - None of the above

11. At present, how confident are you in understanding assessment criteria?
12. At present, how confident are you with understanding academic integrity?
13. At present, how confident are you with understanding the importance of academic integrity?
14. At present, how confident are you in understanding plagiarism?
15. At present, how confident are you in understanding the consequences of plagiarism?
16. At present, how confident are you with referencing?
17. At this stage, is there anything connected to academic integrity, e.g. referencing – that you would like more information and guidance on?
18. What type of study resources do you use/are you most likely to use to develop your understanding of academic integrity, including referencing?
19. Please list any resources that you find helpful with referencing:
20. How do you think the university could better inform and support students to understand academic integrity and referencing?

Appendix 2: Online Staff Survey Questions

1. The number of academic offences relating to plagiarism has increased across the university. What do you attribute to this increase?

2. In principle, do you think all students should have opportunities to submit formative assessments?
3. Please state which level of students you teach:
    - Undergraduate
    - Postgraduate
4. Are your students assessed through academic writing?
5. Are there opportunities for your students to submit formative written assessments?
6. Do you provide your students with personalised constructive feedback on formative assessments?
7. This space is for you to add any comments on your experience(s) as a member of teaching staff relating to the practicalities of formative assessments.

Appendix 3: Student Focus Group Discussion Points

Academic Integrity – Plagiarism and Referencing

- Show of hands: how many of you understood the importance of academic integrity before commencing your course?
- Show of hands: how many of you understand the importance of academic integrity at this stage?
- Show of hands: how many of you feel comfortable and confident in your ability to reference now?
- At present, what does **academic integrity** mean to you?
- What does **academic practice** mean to you?
- Can you remember the context in which you first began to hear discussions on academic integrity at MB University? (Where? When? Who/what department was delivering the message/information?)
- How did that make you feel?
- In groups, please discuss the contexts in which you have heard about academic integrity, referencing and plagiarism. – The context? How did the communication make you feel? Was it clear?

Resources Available

- What (if anything) have you struggled with/found difficult with connected with academic integrity? (e.g. understanding when to reference/the mechanics of referencing/understanding what plagiarism is) *What is it that you find challenging?
- What resources do you find useful in helping you better understand academic integrity/referencing and plagiarism?
    - What exactly do you use these resources for?
    - What features of these resources do you most like/find most helpful?
    - What is unhelpful in this type of resource? (why?)

- How would you advise MB University to better inform and support students in their understanding of assessment criteria, academic integrity – particularly understanding of referencing and plagiarism?
- The LDC provides 1-1 advice, online resources (BB), webinars (open programme) and embedded taught sessions. – How have you found these services? Do you have any comments?
- How could the LDC improve how it supports students better understand academic integrity?

Academic Regulations

- Where do you find information about assessment criteria?
- Do you feel that you understand the rules and regulations of assessment at MB University? – Including what is considered an academic offence? The penalties?
- Do you know where to find out more about academic regulations – i.e. rules, academic offences?

University Comms on Academic Misconduct

- Please look at these documents and think about the messaging (templates informing students of their academic offence and inviting them to discuss the offence).
- What do you think the purpose of this message is?
- What do you think about the clarity of the messages?
- How would you feel if you received these messages? – Why?
- How could these messages be improved?

Turnitin

- What is Turnitin?
- What is your understanding of how MB University uses Turnitin?
- What do you think Turnitin reports reveal/tell the assessor about your writing?
- Do you feel confident in your ability to interpret you Turnitin reports? (why/why not?)
- Are you aware of where you can find guidance on Turnitin reports?
- In some universities, students can submit their assessments through Turnitin and view their Turnitin reports before their assessment deadline. What do you think about this approach?
- If you did have access to Turnitin – how would you use it?
- *What is your awareness, experience and views towards Chat GPT and quillbot – or other similar resources?*
- Do you have any other comments that you would like to add? Final thoughts? Points that you think the university could learn from to better support students with their understanding of academic integrity and to improve academic practice?

Appendix 4: Semi-Structured Interview Schedule

- The number of academic offences is increasing. Why do you think that is?
- What do you think are the key challenges/issues that students face regarding understanding academic integrity? – And – what (if anything) underpins these issues?
- How do you support your students/What steps do you take to ensure that your students receive sufficient information and support understanding and developing concepts relating to academic practice, including academic integrity?
  - How could this be developed?
- Are you aware of any of the following: essay mills, chatgpt and/or quillbot? – If so, what are your thoughts and experiences with them?
  - Do you think these are used by our students? – if so, how?
  - What are your thoughts on the use of these 'resources' by students?
- From a L&T perspective, what are our responsibilities surrounding these resources?
- What can we do (as a university) to reduce/avoid and/or overcome issues relating to the use of these resources?
- How do you think the University should approach conversations about ChatGPT? i.e. what should the University's stance and practice be?
- What do you think is the best strategy/approach that the university could take to support students improve their academic practice, including understanding of academic integrity?
- What do you think is working in terms of support/resource(s) available from the LDC in supporting students and the college in this area?
  - How could this/can these be developed?
- What do you think of peer-peer feedback/support as a means of helping students develop their understand academic integrity? (please elaborate – i.e. why/why not – how/what context?)
- Academic offences – do you feel you have an understanding of the regulations and process of dealing with academic misconduct? (how would you rate your understanding?)
  - How did you learn about this?
  - Do you think the dissemination and guidance on this type of information could/should be developed for staff? (if so, how)
  - Do you discuss academic offences/regulations in your modules? (when/how)
- Do you currently give your students access to Turnitin before final submissions?
- How confident are you in your ability to interpret Turnitin reports? – where did you learn about this?

- Do you think that MB University has helped you develop your understanding of Turnitin reports? (how/where)
- Do you feel you require further support in understanding Turnitin and supporting students in their understanding of Turnitin reports? (if so – how could this be provided?)
- How could the university better support students and staff in their understanding of using Turnitin?

References

Adler, P. A., & Adler, P. (1987). *Membership roles in field research*. Sage.

ALDinHE. (2022). Association for Learning Development in Higher Education. Available at: http://aldinhe.ac.uk/. Accessed 09/09/22.

Ashford-Rowe, K., Herrington, J., & Brown, C. (2013). Establishing the critical elements that determine authentic assessment. *Assessment & Evaluation in Higher Education, 39*(2), 205–222. https://doi.org/10.1080/02602938.2013.819566

Bloor, M., Frankland, J., Thomas, M., & Robson, K. (2001). *Focus groups in social research*. Sage.

Braun, V., & Clarke, V. (2022). *Thematic analysis: A practical guide* (eBook ed.). Sage. https://read.kortext.com/reader/epub/1069943

Brinkmann, S., & Kvale, S. (2015). *Interviews: Learning the craft of qualitative research interviewing* (3rd ed.). Sage.

Carroll, J., & Ryan, J. (2005). *Teaching international students: Improving learning for all*. SEDA.

Case, C. J., King, D. L., & Case, J. A. (2019). E-cheating and undergraduate business students: Trends and the role of gender. *Journal of Business and Behavioural Sciences, 31*(1), 102–113. https://doi.org/10.1007/s10734-009-9274-1

Creswell, J. W. (2009). *Research design: Qualitative, quantitative, and mixed methods approaches* (3rd ed.). Sage.

Dickenson, J. (2023). 'It looks like you're trying to assess a student.' *WonkHE*. 29th March. Available at: https://wonkhe.com/wonk-corner/it-looks-like-youre-trying-to-assess-a-student/

Draper, M., & Fisher, D. (2020). A student and staff partnership in the development of a student directed independent learning toolkit. *International Journal for Students as Partners, 4*(2), 90–96. https://doi.org/10.15173/ijsap.v4i2.4086

Guo, B., Zhang, X., Wang, Z. Jiang, M. Nie, J., Ding, Y., Yue, J. & Wu, Y. (2023). *How close is ChatGPT to human experts? Comparison corpus, evaluation and detection* (Cornell University, Working Paper). Available at: https://arxiv.org/abs/2301.07597. Accessed 30/1/23.

Hennink, M. (2014). *Understanding focus group discussions: Understanding qualitative research*. Oxford University Press.

Krügel, S., Ostermaier, A., & Uhl, M. (2023). *The moral authority of ChatGPT* (Cornell University Working Paper). Available at: https://arxiv.org/abs/2301.07098. Accessed on 30/1/23.

Liles, L. A. (2019). Plagiarism in higher education. *Voices in Education, 5*, 29–32. https://www.college.bm/images/documents/bcj/Vol_5/Plagiarism_in_Higher_Education.pdf

Livingston, K., & Ling, L. (2022). Third space workers in higher education in times of dislocated complexity. *European Journal of Education, 57*, 646–658. https://doi.org/10.1111/ejed.12523

McMahon, P. (2018). 'Making the grade': A grounded theory explaining the student experience of Asian and middle eastern postgraduates in a British university. *Journal of Higher Education Policy & Management, 40*(1), 34–47. https://doi.org/10.1080/1360080X.2017.1411061

Mphahlele, A., & McKenna, S. (2019). The use of Turnitin in higher education: Decoding the myth. *Assessment & Evaluation in Higher Education, 44*(7), 1079–1089. https://doi.org/10.1080/02602938.2019.1573971

Mullholland, M. (2020). Honour and shame: Plagiarism and governing student morality. *Journal of College and Character, 21*(2), 104–115. https://doi.org/10.1080/2194587X.2020.1741394
Noorbehbahani, F., Mohammadi, A., & Aminazadeh, M. (2022). A systematic review of research on cheating in online exams from 2010 to 2021. *Education and Information Technologies, 27*, 8413–8460. https://doi.org/10.1007/s10639-022-10927-7
Pecorari, D. (2010). Academic writing and plagiarism: A linguistic analysis. Continuum.
QAA. (2022). *Revisiting academic practice from a student perspective.* Accessed from: https://www.qaa.ac.uk/docs/qaa/news/quality-compass-revisiting-academic-integrity-from-a-student-perspective.pdf. Accessed on 1/8/23
QAA. (2023a). *Maintaining quality and standards in the ChatGPT era: QAA advice on the opportunities and challenges posed by Generative Artificial Intelligence.* Available at: https://www.qaa.ac.uk/docs/qaa/members/maintaining-quality-and-standards-in-the-chatgpt-era.pdf?sfvrsn=2408aa81_10. Accessed on 14/8/23.
QAA. (2023b). *Reconsidering assessment for the ChatGPT era: QAA advice on developing sustainable assessment strategies.* Available at: https://www.qaa.ac.uk/docs/qaa/members/reconsidering-assessment-for-the-chat-gpt-era.pdf?sfvrsn=38d3af81_6. Accessed on 14/8/23.
Russell Group. (2023). *Russell Group principles on the use of generative AI tools in education.* https://russellgroup.ac.uk/media/6137/rg_ai_principles-final.pdf.
Silverman, D. (2016). *Qualitative research* (4th ed.). Sage.
Staton, B. (2023). Universities express doubt to detect AI-powered plagiarism. *Financial Times.* 3rd April. Accessed from: https://www.ft.com/content/d872d65d-dfd0-40b3-8db9-a17fea20c60c.
Whitchurch, C. (2015). The rise of the third space professionals: Paradoxes and dilemmas. In U. Teichler & W. K. Cummings (Eds.), *Forming, recruiting and managing the academic profession* (pp. 79–99). Springer. https://doi.org/10.1007/978-3-319-16080-1_5
Williamson, K., & Johansen, G. (2018). *Research methods: Information, systems and contexts.* Chandos Publishing.
Yazici, S., Durak, H. Y., Dünya, B. A., & Sentürk, B. (2023). Online versus face-to-face cheating: The prevalence of cheating behaviours during the pandemic compared to the pre-pandemic among Turkish University students. *Journal of Computer Assisted Learning, 39*, 231–254. https://doi.org/10.1111/jcal.12743
Yin, R. K. (2009). *Case study research; design and methods* (4th ed.). Sage.

Chapter 11
Creating Guidance on Appropriate Use of AI Tools in Relation to Students' Academic Work

Irene Glendinning

Abstract In March 2022 concerns were raised to the Academic Integrity Steering Group (AISG) at Coventry University Group, UK about noticeable improvements in the quality of outputs from Google Translate, QuillBot, Grammarly and similar tools. A working group was established to explore recent technological advances, and to develop guidance for staff and students about appropriate and inappropriate uses of AI-powered tools. In September 2022, after exploring literature and inviting experts to brief the working group, the institutional regulations were modified, adding statements about misuse of AI tools. A consultation with University staff and students started with centrally organised staff development workshops from early November 2022. Then, on 30th November 2022, ChatGPT was released.

Between January and July 2023, a total of 25 interviews and focus groups took place, involving 114 staff and students, to share the knowledge gained by the working group and to capture viewpoints and ideas for the guidance. Findings from the consultation helped to frame the guidance notes and resources for students and staff, starting with brief student guidance released in May 2023, with more detailed guidance for staff and students released in September 2023. However, guidance is still being developed more than 1 year on.

The chapter explains the journey followed, what emerged, and lessons learnt, so that others can benefit from this experience.

Keywords Artificial intelligence · Generative AI · Transformative AI · Academic integrity · AI guidelines · Institutional policies

---

I. Glendinning (✉)
Independent Researcher, Leicestershire, UK
e-mail: ireneglendinning@googlemail.com

M. Davis, C. Baptista (eds.), *Ethics and Integrity in Education (Practice)*, Ethics and Integrity in Educational Contexts 10,
https://doi.org/10.1007/978-3-031-95686-7_11

Introduction

The author of this chapter is the former Academic Integrity Lead for Coventry University Group, with many years' experience in developing and maintaining policies for academic integrity, promoting good practice and dealing with a broad range of academic misconduct cases. The main campus of the institution is situated in the West Midlands region of the United Kingdom (UK); it gained university status in 1992 and was formerly a polytechnic. The institution has a richly diverse population of students and staff, with over 40,000 students studying across several campuses both in the UK and overseas.

Early in 2022 the Academic Integrity Steering Group (AISG) identified the need to develop clear guidance for students on appropriate uses of paraphrasing tools, translation tools and essay bots, particularly relating to international students and those with special learning needs. In addition, reports from academic conduct officers in the institution suggested that students misuse of these tools was increasing, for example paraphrasing tools like Grammarly and, more recently the tool QuillBot (n.d.), were being used by a small number of students to disguise plagiarism. There were also cases where international students were first writing all or most of their assignment in their own language and then using translation tools, such as Google Translate and DeepL, to convert into English before submitting. At that time, the outputs from these tools were far from perfect, and often included serious errors in grammar and vocabulary, including inappropriate synonyms or "tortured phrases" (Cabanac et al., 2022a, 2022b; Else, 2021), making it possible for tutors to detect and report these types of inappropriate conduct. Prior to this study the institution had no policies or guidance on these types of conduct, therefore students could legitimately claim that they were not aware this was unacceptable academic practice.

At this time, as a result of advances in AI technology, marked differences were observed by the author and colleagues in the quality of the outputs being generated by these tools. We had already developed comprehensive guidance for students on topics such as the use of proof-readers, contract cheating or the use of a third party to write an assignment, but there was no available institutional guidance on the use of AI-based tools. Accordingly, an AISG working group was established in March 2022, to bring together a range of staff and student representatives. The aim of the working group was to develop institutional guidance on AI use for staff and students. The objectives were to (1) familiarise ourselves with the range and types of tools with AI capabilities that could be used by staff and students; (2) share what we learn with colleagues and students across the institution; (3) capture ideas, views and experiences from within our community, staff and students, about both legitimate and inappropriate uses of AI tools by students and staff; (4) explore ways of detecting and evidencing misuse of AI tools by students, especially in credit-bearing assignments; (5) and to develop guidance about the need to educate staff and students on positive uses of AI in student assignments. The methods we used to achieve these aims and objectives will now be discussed.

Methodology

The working group, led by the author, started working on our responses to AI in March 2022. This predated the release of ChatGPT and the wider interest in this topic. We established from the outset that the results of our study could be useful to other researchers and institutions and aimed to publish our findings. Therefore, we put together our research plan and submitted this for ethical clearance by the University.

To meet the first objective, during the summer of 2022 we invited two experts on this topic, Dr. Thomas Lancaster from Imperial College, London and Dr. Robin Crockett, from University of Northampton, to present their perspectives on AI to the working group and to answer questions. Experts within our own community also contributed to the acquisition of knowledge by sharing their experiences and knowledge with other members of our working group.

Working group members tried a range of free AI tools, including paraphrasing tools, translators, essay bots, to gain first-hand experience of using them and to assess the characteristics and quality of their outputs. A series of internal staff development sessions were organised to keep the University's wider community informed about the developments and to invite colleagues and students to contribute to the consultation process.

By August 2022 we had amassed enough information to incorporate the first changes to institutional regulations, for release in September 2022. The definition of academic misconduct in the institutional regulations was completely rewritten, both to clarify and to extend the types of conduct that needed to be explicitly included in the definition. Two of the 20 items in the new list were specifically about AI. The wording of both of these items was slightly modified one year later in the light of developments and feedback, resulting in these statements:

Academic misconduct includes ethical misconduct and is defined as:

(a) *any attempt to gain an unfair advantage in an assessment, examination or test. This includes (but is not confined to):*

   (i) Unauthorised and unacknowledged use of technological aids and artificial intelligence, including translation software, paraphrasing tools, text generation software (essay bots) or tools to generate programming code, graphics and artwork;

   (ii) Where an assignment is required to be written in English, writing it in another language then using translation software or assistance from a third party to convert into English.

   (Coventry University Group, 2023).

In addition to defining types of integrity breaches, an urgent need was emerging for clear guidance on how, when and what types of AI tools could be used by students and staff in positive and beneficial ways. It was clear that many different AI tools can perform a wide range of specialist generative and transformative functions. It was important to include and discuss the different types, which led to the categorisation in Fig. 11.1. An earlier version of this figure was used in the initial briefings and development sessions for staff starting in November 2022 and contributes to the essential structure of the guidance notes for staff and students that were developed following the consultation.

The consultation involved several elements. Colleagues from across all parts of our institution were invited to focus groups. Individual meetings were arranged for senior colleagues from the institutional leadership team. The Student Union was tasked with organising and running a series of focus groups with students. In addition, a short questionnaire was designed to supplement the data collection and allow for information and views to be shared with us that individuals did not wish to raise during a focus group. The questionnaire also made it possible for people unable to join the focus groups and interviews to contribute their views.

The consultation was designed as a two-way process. Focus group and individual meeting facilitators, who were all members of the working group, first provided the same 10–15-min presentation for participants with information collected by the working group about AI tools. This helped to set the scene for the discussions and provided opportunities to answer any immediate questions from participants. A series of standard prompts were available as a framework for the discussions, with flexibility to adapt to accommodate different levels of experience of specific participants.

On 30th November 2022, when ChatGPT was released, (OpenAI, 2022), the working group had 9 months' start in developing guidance compared to most institutions. This momentous development meant that we had an increasingly willing and highly interested set of participants for our consultation. Focus groups and interviews were held between January and July 2023, with one late focus group on 3rd October 2023. Most of the focus groups were held face-to-face, a few were hybrid meetings, and some were conducted fully online. Participant responses and discussions were mainly recorded by manual note taking during the meetings, but a few sessions were audio recorded, with permission from participants, mainly when

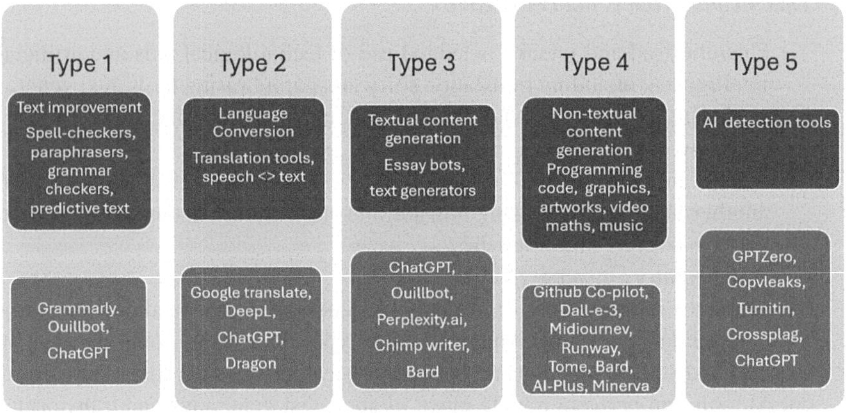

Fig. 11.1 Classification of artificial intelligence tools

Table 11.1 Summary of responses to consultation

Type of meeting	Number of events	Number of participants
Staff focus group	15	97
Senior staff meeting	8	9
Student focus group	2	7
Totals	25	114

a note-taker was not available. The main points raised in these meetings were transcribed later. Table 11.1 provides a summary of the consultation involving 114 participants in total with 25 consultation events.

In addition to consultation meetings, 19 questionnaire responses were collected from staff and one response from a student, with all 20 respondents indicating that they had participated in the consultation. Participants included academics from every University faculty and campus, librarians, academic writing tutors, educational technologists, learning designers, student union advisors, student representatives, student support service staff and elected sabbatical (student) officers. The number of student participants was disappointingly low, but their feedback was very useful.

A thematic analysis of the meeting feedback was conducted to extract the main themes raised by participants. The small number of questionnaire responses meant that the analysis of this data could be conducted manually.

The results and details of how we applied these findings are detailed in Section "Findings from Institutional Consultation on Perspectives on Students' Use of AI Tools in Relation to their Assessments" of this chapter. Next is a summary of the main literature sources that helped to guide us in the development of our institutional approach and policies.

Literature Review

This literature review is presented chronologically to demonstrate how the literature and guidance evolved over a very short period of time to reflect the way the capabilities of AI tools evolved during that time.

One of the earliest sources we found, published early in 2021, concerned the use of QuillBot (n.d.) by students learning English for improving their grammar, summarising, paraphrasing writing, and hiding or disguising plagiarised text, rather than taking time to practise to develop "higher-order thinking as well as strong reading and writing skills" (Fitria, 2021, p.184). Worryingly, the article concludes by actively advocating the use of QuillBot by students, rather than pointing out the impacts that it would have on language students' learning. Interestingly, QuillBot was the AI tool used in the first successfully proven AI misuse case the author was aware of at the institution, in October 2022.

Another early paper relevant to our study, published in March 2021, discussed use of "machine translation" tools by students studying English for academic purposes (Groves & Mundt, 2021). This paper is especially relevant because it is a pre-ChatGPT study, based on interviews, of how academic staff in two UK universities viewed the use of freely available tools like Google Translate by students studying in UK higher education whose first language is not English (Groves & Mundt, 2021), which called for a broader conversation across the UK HE Sector about institutional policies relating to tools of this type.

An academic paper published in January 2022 focused on paraphrasing by AI with machine-generated synonyms as a means of fraud in academic publishing (Cabanac et al., 2022a, 2022b), although the same principles apply to use or abuse of AI by students. At this time unchecked synonym replacement was relatively easy to detect because of strange expressions, for example "artificial intelligence" became "counterfeit consciousness" (Else, 2021). However, ongoing advances in GenAI make it increasingly difficult to identify and evidence this type of misuse.

The global popular press and media had a field-day after ChatGPT was released, with many articles forecasting the end of education as we know it (Cuthbertson, 2020) and rampant cheating by students (Whitford, 2022). Several countries and institutions announced a ban on student use of AI tools, including the New York City school system (The Guardian, 2023) and the Australian states of New South Wales and Queensland (Johnson, 2023). The release of ChatGPT served as a catalyst for universities around the world to put policies in place, and many struggled to know how to respond, as illustrated by this quotation from the report on a 2023 study of 1250 UK undergraduate students conducted for the Higher Education Policy Institute (HEPI): "*While students are quite confident that staff understood their AI use for assessments, they are much less positive about how prepared staff are to help them work with generative AI. Under a fifth of students (18%) agree staff are well-equipped, with many more students (29%) disagreeing*" (Freeman, 2024, p.10).

In April 2021 Jisc had published a report for the UK HE Sector with general guidance on Artificial Intelligence, which was subsequently updated twice in response to the rapid changes (Jisc, 2023). Jisc is a digital leader for the UK HE Sector, which has been regularly offering key insights into technological developments. In Oct 2021, its focus was on how ready institutions were for AI and how it might fit with institutional strategies and policies (Jisc, 2021). By Sept 2023 there was an acknowledgement by Jisc that AI cannot be ignored or avoided and general advice as to how to manage its use, acknowledging the many ethical challenges now faced in education, which included these five principles set out in a statement from the Russell Group:

- Universities will support students and staff to become AI-literate.
- Staff should be equipped to support students to use generative AI tools effectively and appropriately in their learning experience.
- Universities will adapt teaching and assessment to incorporate the ethical use of generative AI and support equal access.

- Universities will ensure academic rigour and integrity is upheld.
- Universities will work collaboratively to share best practice as the technology and its application in education evolves.

- (Russell Group, 2023; Jisc, 2023, p.20).

Jisc also published a second report in 2022 specifically focusing on Generative AI, with the third edition of the same guidance document released in January 2024 (Jisc, 2024). The Jisc guidance recommends appropriate uses while acknowledging that there is little in UK law to regulate AI development at present (Jisc, 2023, 2024).

Guidance on ethical use of AI by students began to emerge early in 2023, for example (Eaton & Anselmo, 2023; Foltýnek et al., 2023). The UK's Quality Assurance Agency QAA produced their first guidance for their member institutions on AI in January 2023 (QAA, n.d., 2023a). Further guidance followed during 2023 from QAA and other national and international agencies concerned with quality and integrity (QAA, 2023b; QQI, 2023; NAIN, 2023; UNESCO, 2023). The main messages in the guidance included acceptance that AI will continue to be integral in the modern world and there is a need to design assessments that cannot simply be answered by an AI. As a result of these comprehensive and balanced recommendations, more nuanced statements and guidance came from institutions and mission groups (including: University of Sydney (Liu & Bridgeman, 2023, June 8); Russell Group, 2023).

In addition to the many press and media articles, and academic publications and reports, numerous webinars were organised, and this became the primary topic at many conferences and networking events focused on educational research, including several workshops and presentations by the members of the AI working group I had established (NERQ, 2023; ECEIA, 2023; EATAW, 2023).

One major concern expressed by participants at these events was the difficulty of detecting misuse of artificial intelligence tools by students in their assessed work. Indeed, the author encountered many people, both from my own research and in my experience of attending academic conferences and events, who believed that detection of AI misuse by students was already impossible. At the end of January 2023 OpenAI launched a 'classifier trained to distinguish AI-written and human-written text' (OpenAI, 2023, January 31).

An announcement was made by Turnitin on 28th February 2023 that they would upgrade their software to add "an AI detection tool", for all customers on 4th April 2023 (Turnitin, 2023, April 4). Discussions across the UK's well-connected HE sector about the lack of advance warning about this change, combined with providers having no control of the detection tool (Knox, 2023), led to a formal expression of concern addressed to Turnitin's product manager, representing four UK organisations: UCISA, the Heads of E-Learning Forum (HeLF), Jisc and APUC. Following this communication, a total of 153 UK Turnitin customers, mainly universities, chose to opt out of the upgrade (for example, University of Greenwich, 2023, March 31). The Turnitin AI detection tool was later made available to all customers, complete with essential controls, allowing institutions to test the software before deciding whether to give access and to whom, which led to decisions in other parts of the world to disable the AI detection tool, for example Vanderbilt University (Coley, 2023, August 16).

OpenAI's classifier and the Turnitin detection tool were not the only AI detection tools available. The detectors GPTZero (n.d.) and Copyleaks (n.d.) both became available for free from January 2023. In addition, it was discovered that ChatGPT and similar chatbots could be used to detect any AI generated text (Khatsenkova, 2023). However, through both institutional testing and more formalised research it soon became apparent that the detection capabilities of all the available tools were not accurate or reliable enough to detect or evidence AI use (Weber-Wulff et al., 2023; Crockett, 2023). Only 6 months after it was released, OpenAI withdrew its classification tool due to acknowledged inaccuracies and lack of reliability (Kirchner et al., 2023). As a result of the testing, serious questions were highlighted about copyright, data protection and data privacy of any of any materials uploaded to chatbots and detection tools (Foltýnek et al., 2023).

Despite the available research and guidance, cases continue to emerge of institutions blindly relying on outputs from AI detection tools as evidence of AI misuse by students, including the UK's Birkbeck (Topinka, 2024), leading to false and inappropriate allegations against students.

As this chapter largely focuses on academic integrity and AI, literature sources promoting positive uses of AI have not been included here. However, an essential element of institutional policies and guidance is understanding what additional skills students and staff need to acquire to benefit from AI in their study, learning, work, personal life and future careers. For example: Pre ChatGPT, AI was used to great effect in predicting protein shapes and their properties (Callaway, 2022), which could have taken many decades to complete if done by humans.

On the same theme, one final source to note is an excellent recently published survey of students at the University of Northampton, UK (Lea et al., 2024). The Northampton team conducted questionnaires in 2023 and 2024 to capture information about ways that students are using GenAI and their views on AI use in education (Lea et al., 2024), comparing both sets of responses in the report. The results are well worth exploring in more detail, but the findings this author found remarkable are (1) the percentage of student respondents who were "non-users" of AI (approximately 50% in 2023, almost 60% in 2024), with some of them clearly seeing AI as a great threat to education (Lea et al., 2024, p.6, p.11); (2) the marked difference in viewpoints between UK and international students on value of using of AI - almost 60% of international students expressed positivity about "GenAI advancements", compared to about 5% of UK students (Lea et al., 2024, p.8). These findings bring important messages to those developing policies and guidance, especially the need to understand the range of different student experiences relating to use and abuse of AI, in all its forms.

As the field of knowledge regarding the uses of AI tools has grown so quickly, some of the sources included in this review were first released as preprints, then later published in peer reviewed journals. Clearly, this chapter includes only a small selection of the recent literature.

Findings from Institutional Consultation on Perspectives on Students' Use of AI Tools in Relation to their Assessments

The consultation proved to be an effective approach, not just for collecting participants' knowledge, views, requirements and ideas, but also for communicating what the working group had learned so far and for sharing our initial ideas for policies and guidance. The feedback helped to refine our ideas.

One result was the variation in levels of knowledge amongst staff (see Table 11.2). Translation software was one AI tool that more staff felt they had good knowledge of (11/20), with grammar checking tools felt to be the next best understood, (9/20) followed by grammar checking and text generation software. However, this still leaves nearly half the staff feeling that they have very limited knowledge of all types of AI. This had implications for the creation of policy and guidance. This then also impacted how likely they were to have used the tools in some context with students.

Other groups around our own campus were identified who were working on different aspects of AI, which allowed us to pool our ideas, resources and materials. For example: a team of learning technologists and developers from the University's online learning group had already conducted an evaluation of AI detection tools; Arts and Humanities faculty were exploring AI tools and outputs other than generated text; and Computer Science had already developed policies for utilising Github Desktop to record students' own work and their use of AI. The experience of participants was mixed, ranging from staff classifying themselves as having no knowledge to staff classifying themselves as having expert knowledge. The analysis of the 20 questionnaire respondents in Table 11.2 helps to illustrate this range of perceived expertise levels.

The analysis of the qualitative data took much more time and effort. A manual reading of the notes from all 25 interviews and focus groups identified a range of themes and sub-themes, which were further grouped to produce the list of topics in Table 11.3 as a basis for structuring the staff and student guidance notes.

The first drafts of guidance notes and materials for staff and students were designed during May 2023, structured according to the thematic analysis. They were circulated for comment to a wide range of people, both working group members, student representatives and participants in the consultation. After all concerns and comments were addressed, the materials were made available to staff via the University's Teaching and Learning Knowledge Base, to be used during induction for new students joining the University at that time.

After the first draft guidance notes were circulated, further refinement continued throughout the summer of 2023, in response to feedback and new information. A hybrid event was held on 12th June 2023 on academic conduct and integrity, where the initial findings and ideas were presented to an in-person audience of about 70 staff and student representatives and approximately 20 on-line participants. The first sets of guidance for staff and students were released in September 2023 on the internal web portals for staff and students, as part of the still on-going awareness-raising

Table 11.2 Questionnaire participants' perceived knowledge of AI tools

Knowledge	Translation	Paraphrasing	Grammar check	Text generation	Smart devices	Other tech	Graphics	Code gen	Music gen
None	1	6	3	5	14	15	13	15	14
Very little	1	0	0	0	1	0	0	0	0
Limited	6	8	8	6	3	5	3	1	4
Reasonable	1	0	0	0	0	0	0	0	0
Good	9	6	7	6	0	0	2	2	1
User	1	0	0	1	0	0	1	0	1
Expert	0	0	0	1	1	0	1	1	0
Other	1	0	2	1	1	0	0	1	0
#	20	20	20	20	20	20	20	20	20

Table 11.3 Thematic analysis, themes and sub-themes

Theme	Sub-themes
Appropriate uses of AI tools	What good use of AI looks like When to permit use Examples of innovative and appropriate use of AI tools
Misuse of AI tools	– Examples of risks from AI tools – When and how to restrict use – Admissions and recruitment
Guidance for staff	– Teaching, learning, pedagogy, – Assessment methods, assessment design – Workplace use of AI tools, future careers – Group work – Staff training and research
External contacts	– Supporting partner institutions and schools to develop policies on AI
Academic conduct process	– Challenges and disincentives for staff to report cases – Detecting potential misuse – Use of AI detectors – Understanding reasons for cheating – Managing risks – Regulations, classifying seriousness of misuse, – Quality assurance
Deterring	– Guidance for students – Educating students about AI tools – Educating students about academic integrity – Developing student skills
Equality and diversity	– Use of AI tools to support students and staff with specific learning difficulties – Inclusivity and AI – ensuring fair and equal access to AI for all students.

campaign involving training for staff and students that began during the autumn semester of 2023.

Throughout the academic year of 2023–2024 regular centrally organised staff development sessions were run, which generated useful feedback on ways to improve the guidance and facilitate the acknowledgement process for legitimate use of AI tools. Guidance sessions for students were organised centrally by the Student Union and also through tutorials at course/programme level.

As part of the on-going awareness-raising, a day-long on-line event on AI was organised for University staff and student representatives on 13th June 2024, which allowed the expertise from external sources and some of the wider working group members to be shared with over 100 participants. The morning was devoted to receiving valuable input from Thomas Lancaster (Imperial College, London), Robin Crockett (University of Northampton) and Mary Davis (Oxford Brookes University). The afternoon consisted of six 20-minute interactive presentations from experts within the University: AI in mathematics and statistics; Developing online content with AI; What is GenAI; Developing a workshop to support students' use of AI in academic writing; Academic Integrity and academic conduct; Curriculum development in the age of AI. Panel sessions in the morning and the afternoon both generated lively discussions. The day's sessions were recorded for the benefit of those unable to attend and those wishing to review the presentations.

The chapter will now consider current and future challenges and how these are being addressed.

Discussion

The Main Challenges

The findings from the focus groups and questionnaires reflected the guidance coming from across the world, including Jisc and QAA, that simply banning AI use was not an adequate response, but defining appropriate use proved far more complex.

Introducing the requirement for students to seek permission to use specific tools for specific purposes required clear guidance for staff on when to give permission and how and where to articulate this, to ensure clarity and consistency in implementation.

Many AI tools are already used by people with specific learning difficulties as part of their agreed support arrangements, therefore the guidance and related submission methods needed to allow for appropriate exemptions.

Any policy needed to be as future-proof as possible to ensure it remained relevant as the technology improves, for example many of the features of Type 1 tools (Text Improvement) and Type 2 tools (Text Transformation) are already embedded in commonly used software tools, making it unrealistic to ask students to acknowledge their use, any more than they would declare they had used a word processor.

The guidance is potentially more important than any formal policy. Guidance needed to cover several topics. Firstly, explanations of different AI tools and their functions for those unfamiliar with them, which, based on the focus groups, could be almost 50% of staff. Secondly, the guidance needed to provide exemplars of use across a range of subjects and task types. Thirdly, the guidance had to address the process by which the use of AI could be clearly described in students' assignments. Fourthly, the application needed to be straightforward and easy to use for everyone.

Every member of the institutional community needed to be equipped with skills and knowledge of what AI tools can do and how they can use AI tools to help their own productivity, to support their work and study. This required the design of a comprehensive training programme for everyone. Fortunately, the wider working group included several people with specialist skills in designing and delivering suitable materials for staff development. This educational task is now starting, but educating the whole community remains a long-term mission.

Academic staff urgently needed guidance, not just on what constitutes appropriate and inappropriate uses of AI tools, but also on how to identify cases of misuse and what evidence would constitute sufficient proof of this. Therefore, guidance was developed for both academic staff and academic conduct officers (ACOs) relating to these aspects, including a checklist of characteristics of AI generated or transformed content (Glendinning, 2024) and guidance for ACOs on the level of seriousness associated with different types of conduct.

Most of the discussions so far in this chapter have focused on text-based written work, but as illustrated in Fig. 11.1, AI tools can also generate other content, including music (Artificial Intelligence, 2023), videos and PowerPoint (Tome, 2024, August 23), graphics and artwork (Gemini, n.d.; DALL-E 3 (OpenAI, 2024)), computer programming code (OpenAI, 2022, November 30; Github, n.d.). Misuse of AI is also associated with Types 1 and 2 tools that enhance and transform rather than generate content, therefore the guidance and associated policies needed to cover these aspects.

An important purpose of the staff guidance concerns designing assessments that facilitate good engagement and motivation for students to learn and develop knowledge and skills in ethical use of AI tools and to make it difficult for students to outsource their work (Bowditich, 2023, June 14). These types of assignments are largely the same as those recommended to address contract cheating (Dawson, 2020). Therefore, the guidance must include examples of where appropriate and innovative uses of AI can be included in learning experiences and assignments, across subjects and disciplines.

Guidance for students presented further challenges. It needed to be easily accessible, with core messages that could be promoted in a few key points, as well as being attractive and engaging, accessible, easy to understand and as succinct as possible. Guidance was needed both for students to access themselves and for staff to use to educate their students.

Defining Appropriate Student Use

One of the biggest challenges was providing clarity for students and staff on what constitutes acceptable use of AI and what is unethical (Sladdin, 2024, February 8; Spisak et al., 2023, June 30). Most importantly, the guidance and associated forms and declarations had to factor in appropriate use of AI tools by students with special educational needs.

The main principle behind any guidance on which AI tools can and cannot be used and for what purposes depends on the nature of each assessment task. For this reason, it became clear that the specific guidance to students must be provided at the level of individual assessments, therefore must sit within each assessment brief at module or study unit level (for example, see Perkins et al., 2024). To ensure this can happen in a clear and unambiguous manner, very specific templates, guidance and associated training were needed for all academic staff.

In addition to the statement in the regulations about requiring approval, the working group proposed a three-step process called Acknowledge, Describe and Reference. The initial idea was for students to acknowledge at the point of submission that they had used generative AI, they would reference any content from AI and finally, they would use an appendix to describe in detail how they had used AI. Further discussion raised concerns that the process is lengthy and would create a barrier to students, so consideration was given to allow students to use a form to acknowledge and describe their use and included citing and referencing in the normal way.

There are three elements to the proposed student guidance on use of AI in assessed work:

1. A general brief common statement about AI for students on the assessment brief that applies to all assessments
2. A specific statement from the module leader for each assignment clarifying whether AI use is permitted for this task, and if so, what AI can be used and how.
3. A form and submission method for students to declare (and acknowledge) what AI they have used in this assignment.

A common "assessment brief" template was introduced across the Group in September 2023, which included a short statement about academic integrity and AI use. The template was modified in September 2024 to include more comprehensive guidance on AI use.

Detecting and Evidencing Misuse

Although the majority of students in the Group are diligent and keen to learn, the conduct of a small minority of students who resort to deliberate cheating cannot be ignored. Detecting AI misuse and contract cheating is difficult, time-consuming, often not valued and inadequately resourced and supported (Glendinning, 2024). Therefore, to detect, evidence and report potential integrity breaches staff need to invest additional time and effort at busy times when they face tight deadlines to complete their marking. Although they are improving all the time, fortunately, the outputs from GenAI are far from perfect, which means that there are still clues that can help academics to detect misuse, for example, the inclusion of fictitious or irrelevant content and references; the standard and content of a student submission not reflecting their level of knowledge or writing style (Glendinning, 2024).

This author very much supports the view of Ellis and Murdoch (2024), that an institutional strategy should combine student education on academic integrity with methods for assessment security (Dawson, 2020), to allow for different student attitudes ranging from the majority of students who champion and support integrity to those unable or unwilling to engage in their learning.

Three essential requirements for improving detection and evidencing of all forms of academic integrity breaches are (a) training for all staff, including markers and decision-makers on identifying and interpreting evidence; (b) supporting staff by providing resources and expertise in generating and presenting evidence; (c) rewarding and encouraging rather than disadvantaging staff who report cases.

Further Work

In addition to the continuing development and refinement of materials and resources identified already in the chapter, the consultation identified several areas where future efforts should be focused.

A clear statement from the University's senior team would help to clarify the overall institutional strategy on use of AI ideally with some centrally organised educational sessions to guide students and staff on the institution's position.

Supporting other institutions who are educating our potential future students is in the interest of all universities. If universities help their partner colleges and schools to develop their own policies and procedures, especially relating to academic integrity, then they will benefit in the longer term by admitting students who are already prepared for the next stage of their educational journey. To this end, since May 2023 several well-attended online webinars and smaller in-person staff development sessions were delivered by the author for partner institutions within the UK and internationally, about how our own guidance on AI is developing. In addition, the University continues to share useful materials with partner institutions, for adapting to their own use. However, there is potential for University staff to be much more pro-active in supporting and guiding other institutions, both locally-based and more distant.

There is great scope for forging partnerships with students, ideally via the student union and student representatives, on learning to use AI to support different elements of their assignment processes. It is also essential to raise awareness of the negative impacts on students' learning and skills development if they uncritically over-rely on AI in their study and assessment activities.

Conversely, it is essential that awareness is maintained in every subject area about what skills in AI use students will need after they graduate and what AI tools they will need to utilise, so that graduates will be suitably prepared for their professional lives and future careers. The best way to achieve this is by on-going communications with the appropriate professional bodies, commerce and industry contacts, about their requirements and on-going challenges, as their own ideas and methods continue to evolve.

Conclusion

The chapter provides insight into a consultation undertaken at a UK university about students' use of AI in relation to assignments, guidance that was developed and what is still to be done. Even with the advantage of an early start in establishing a working group on AI in March 2022, developing appropriate guidance has been a lengthy and complex process, which is far from finished at the time of writing this paper.

A remarkable side-effect emerging from the advancement of AI is the sudden upsurge in interest in academic integrity. Those of us who have struggled for years to bring attention to our research, especially to the threats and challenges from contract cheating, have found our knowledge and expertise to be very much in demand in the age of AI.

One of the most challenging aspects of AI is the rapid pace of change. The release of ChatGPT represented a fundamental shift in the quality of outputs produced, raising concerns that students would submit work generated by AI that could not be either identified or proved. The generative AI text-based outputs from tools trained on LLMs after GPT 3 are still rapidly improving in quality resulting in AI generated text increasingly resembling human-written text (Crockett, 2023; Adams, 2024). However, despite these continual advances and the early concerns, it is still proving possible, although very time-consuming, to both identify and prove when students use AI tools inappropriately.

The immediate priority and focus now must be on reevaluating the purpose of education and how learning is measured. This requires innovation in pedagogy, learning approaches, assessment design, on-going staff development, working with students as partners and attention to quality, standards and integrity. Attention to these details will help to address the challenges identified here in educational integrity and standards.

Grateful thanks must be acknowledged to all members of the working group and participants in the consultation described in this study, for their contributions to this paper.

References

Adams, R. (2024, June 26). Researchers fool university markers with AI-generated exam papers. The Guardian. https://www.theguardian.com/education/article/2024/jun/26/researchers-fool-university-markers-with-ai-generated-exam-papers.

Artificial Intelligence. (2023, June 14). *12 Apps and tools to make music with artificial intelligence*. https://www.aiplusinfo.com/blog/ai-music-generators/.

Bowditich, I. (2023, September 12). Assessment Menu*: Designing assessment in an AI enabled world.* JISC. https://nationalcentreforai.jiscinvolve.org/wp/2023/09/12/designing-assessment-in-an-ai-enabled-world/.

Cabanac, G., Labbé, C., & Magazinov, A. (2022a, January 13). "Bosom peril" is not "breast cancer": How weird computer-generated phrases help researchers find scientific publishing fraud. Bulletin of the Atomic Scientists. https://thebulletin.org/2022/01/bosom-peril-is-not-breast-cancer-how-weird-computer-generated-phrases-help-researchers-find-scientific-publishing-fraud/.

Cabanac, G., Labbé, C., & Magazinov, A. (2022b). *The 'Problematic Paper Screener' automatically selects suspect publications for post-publication (re)assessment* [Paper presentation]. WCRI 2022: 7th World Conference on Research Integrity, Cape Town, South Africa. *arXiv*:2210.04895. https://doi.org/10.48550/arXiv.2210.04895

Callaway, E. (2022, July 28). 'The entire protein universe': AI predicts shape of nearly every known protein. Nature News. https://www.nature.com/articles/d41586-022-02083-2#:~:text=29%20July%202022-,'The%20entire%20protein%20universe'%3A%20AI%20predicts%20 shape%20of%20nearly,of%20around%20200%20million%20proteins.&text=From%20today%2C%20determining%20the%203D,typing%20in%20a%20Google%20search.

Coley, M. (2023, August 16). Guidance on AI Detection and why we're disabling Turnitin's AI detector. Vanderbilt University. https://www.vanderbilt.edu/brightspace/2023/08/16/guidance-on-ai-detection-and-why-were-disabling-turnitins-ai-detector/.

Copyleaks. (n.d.). *AI detector*. https://copyleaks.com/ai-content-detector.

Coventry University Group. (2023). General regulations, Appendix 1, 2023–24. https://livecoventryac.sharepoint.com/:w:/s/externalstudentdocuments/EfFxSzhxllxEsZ8y2p2fnd0BWp8LdXfjkRyFS5OVp2Y8og?rtime=s_L8rfn53Eg.

Crockett, R. (2023, November 20). Testing the AI detectors part 2. University of Northampton Learning Technology Blog. https://blogs.northampton.ac.uk/learntech/2023/11/20/testing-the-ai-detectors-part-2/.

Cuthbertson, A. (2020, July 27). Elon Musk claims AI will overtake humans 'in less than five years'. *Independent* (United Kingdom). https://www.independent.co.uk/tech/elon-musk-artificial-intelligence-ai-singularity-a9640196.html.

OpenAI. (2024). Dall-E *3*. https://openai.com/dall-e-3.

Dawson, P. (2020). Defending assessment security in a digital world. Routledge. https://doi.org/10.4324/9780429324178.

EATAW. (2023, June 5-7). Writing technology, thinking, and learning. What tutoring, teaching, and learning of writing mean in a digitalized university. Zurich University of Applied Sciences in Winterthur, Switzerland. https://www.zhaw.ch/de/linguistik/institute-zentren/ilc/news-und-events/detailansicht-veranstaltungen/event-news/eataw-conference-2023/.

Eaton, S., & Anselmo, L. (2023, January). Teaching and learning with artificial intelligence apps Taylor Institute. University of Calgary. https://taylorinstitute.ucalgary.ca/teaching-with-AI-apps.

ECEIA. (2023, July 12–14). *9th European conference on ethics and integrity in academia*. University of Derby, UK. https://academicintegrity.eu/conference/european-conference-on-ethics-and-integrity-in-academia-2023/.

Ellis, C., & Murdoch, K. (2024). The educational integrity enforcement pyramid: A new framework for challenging and responding to student cheating. *Assessment & Evaluation in Higher Education*, 1–11. https://doi.org/10.1080/02602938.2024.2329167

Else, H. (2021, August 5, corrected August 10). 'Tortured Phrases' give away fabricated research papers. Nature News. https://www.nature.com/articles/d41586-021-02134-0.

Fitria, T. N. (2021). QuillBot as an online tool: Students' alternative in paraphrasing and rewriting of English writing. *Englisia: Journal of Language, Education, and Humanities, 9*(1), 183–196. https://doi.org/10.22373/ej.v9i1.10233

Foltýnek, T., Bjelobaba, S., Glendinning, I., Khan, Z., Santos, R., & Kravjar, J. (2023). ENAI recommendation on the ethical use of artificial intelligence in education. *International Journal of Educational Integrity, 19*, 12. https://doi.org/10.1007/s40979-023-00133-4

Freeman, J. (2024, February 24). Provide or punish? Student' views on generative AI in higher education. Higher Education Policy Institute. https://www.hepi.ac.uk/2024/02/01/provide-or-punish-students-views-on-generative-ai-in-higher-education/.

Gemini. (n.d.). Supercharge your creativity and productivity. https://gemini.google.com/.

Github. (n.d.). Github Copilot. https://github.com/features/copilot.

Glendinning, I. (2024). Detecting is just the start, deciding whether to take action and generating evidence is harder. In S. Mahmud (Ed.), *Academic integrity in the age of artificial intelligence* (pp. 220–240). IGI Global. https://doi.org/10.4018/979-8-3693-0240-8

GPTZero. (n.d.). More than an AI detector. Preserve what's human. https://gptzero.me/.

Groves, M., & Mundt, K. (2021). A ghostwriter in the machine? Attitudes of academic staff towards machine translation use in internationalised higher education. *Journal of English for Academic Purposes, 50*, 100957. https://doi.org/10.1016/j.jeap.2021.100957

Knox, L. (2023, April 3). Can Turnitin cure higher Ed's AI fever? Inside Higher Ed. https://www.insidehighered.com/news/2023/04/03/turnitins-solution-ai-cheating-raises-faculty-concerns.

Jisc. (2021, November 17). *A Pathway towards responsible, ethical AI.* https://www.jisc.ac.uk/reports/a-pathway-towards-responsible-ethical-ai.

Jisc. (2023, September 6). AI in Tertiary education, Summary of the state of play. https://beta.jisc.ac.uk/reports/artificial-intelligence-in-tertiary-education.

Jisc. (2024, January). *Generative AI – a primer, Version 1.3.* https://repository.jisc.ac.uk/9407/1/Jan_2024_generative_AI_primer-final.pdf.

Johnson, A. (2023, January 31). ChatGPT in schools: Here's where it's banned – and how it could potentially help students. Forbes. https://www.forbes.com/sites/ariannajohnson/2023/01/18/chatgpt-in-schools-heres-where-its-banned-and-how-it-could-potentially-help-students/.

Khatsenkova, S. (2023, January 19). OpenAI's powerful new chatbot, ChatGPT, has got many wondering how they can tell whether a human or an artificial intelligence wrote what they're reading. EuroNews. https://www.euronews.com/next/2023/01/19/chatgpt-is-it-possible-to-detect-ai-generated-text.

Kirchner, J.H., Ahman, L., Aaronson, S., & Leike, J. (2023, July 20). New AI classifier for indicating AI-written text. OpenAI. https://openai.com/blog/new-ai-classifier-for-indicating-ai-written-text.

Lea, K., Byles R., & Howe. R. (2024). *Report: Exploring student perspectives on Generative artificial intelligence tools at the University of Northampton.* https://blogs.northampton.ac.uk/learntech/wp-content/uploads/sites/7/2024/06/AI-student-survey24_11June_T.pdf.

Liu, D., Bridgeman, A. (2023, June 8). *ChatGPT is old news: How do we assess in the age of AI writing co-pilots?* https://educational-innovation.sydney.edu.au/teaching@sydney/chatgpt-is-old-news-how-do-we-assess-in-the-age-of-ai-writing-co-pilots/?fbclid=IwAR1JSRu8bjMo8H8zX5VJl-BjqyArQxikEpM898MVCLPm_cU3vblLgNE7GpQ_aem_AV28MDpP_1nOo4_m4IJF-hy0EyKy3uN2YBFO4BIms8g7O507zESUxe5fznU4W-JE7T9k.

NAIN. (2023). *Generative artificial intelligence (GenAI) Guidelines for Educators.* https://www.qqi.ie/sites/default/files/2023-09/NAIN%20Generative%20AI%20Guidelines%20for%20Educators%202023.pdf.

NERQ. (2023, March 3). Kick-off meeting of the network for education in research quality (NERQ). European Commission, Brussels. https://www.enrio.eu/kick-off-nerq-network/.

Perkins, M., Furze, L., Roe, J., & MacVaugh, J. (2024). The artificial intelligence assessment scale (AIAS): A framework for ethical integration of generative AI in educational assessment. *Journal of UniversityT eaching & Learning Practice, 21*, 6. https://doi.org/10.53761/q3azde36

OpenAI. (2022, November 30). *Introducing ChatGPT.* https://openai.com/index/chatgpt/.

OpenAI. (2023, January 31). *New AI classifier for indicating AI-written text.* https://platform.openai.com/ai-text-classifier.

QAA. (2023a, January 30). *QAA briefs members on artificial Intelligence threat to academic integrity.* https://www.qaa.ac.uk/news-events/news/qaa-briefs-members-on-artificial-intelligence-threat-to-academic-integrity.

QAA. (2023b, May 8). *Maintaining quality and standards in the ChatGPT era: QAA advice on the opportunities and challenges posed by Generative Artificial Intelligence.* https://www.qaa.ac.uk/docs/qaa/members/maintaining-quality-and-standards-in-the-chatgpt-era.pdf.

QAA. (n.d.). *Generative Artificial Intelligence.* https://www.qaa.ac.uk/membership/membership-areas-of-work/generative-artificial-intelligence.

QQI. (2023, September 8). *New framework for investigating academic misconduct.* https://www.qqi.ie/news/new-framework-for-investigating-academic-misconduct-published.

QuillBot. (n.d.). *Your complete writing solution.* https://quillbot.com.

Russell Group. (2023, July 4). *New principles on use of AI in education.* https://russellgroup.ac.uk/news/new-principles-on-use-of-ai-in-education/.

Sladdin, J. (2024, February 8). UK universities 'must provide clarity' over student AI use. Pinsent Masons. https://www.pinsentmasons.com/out-law/news/uk-universities-must-provide-clarity-over-student-ai-use.

Spisak, B., Rosenberg, L.B., & Beilby, M. (2023, June 30). Principles for using AI responsibly. Harvard Business Review. https://hbr.org/2023/06/13-principles-for-using-ai-responsibly.

The Guardian. (2023, January 6). *New York City schools ban AI chatbot that writes essays and answers prompts*. https://www.theguardian.com/us-news/2023/jan/06/new-york-city-schools-ban-ai-chatbot-chatgpt.

Tome. (2024, August 23). Tome's next chapter. https://tome.app/manifesto.

Topinka, R. (2024, February 13). The software says my student cheated using AI. They say they are innocent. Who do I believe? *The Guardian*. https://www.theguardian.com/commentisfree/2024/feb/13/software-student-cheated-combat-ai.

Turnitin. (2023, April 4). The Launch of Turnitin's AI writing detector and the road ahead. Turnitin Blog. https://www.turnitin.com/blog/the-launch-of-turnitins-ai-writing-detector-and-the-road-ahead.

UNESCO. (2023). *ChatGPT and artificial intelligence in higher education – quick start guide*. https://www.iesalc.unesco.org/wp-content/uploads/2023/04/ChatGPT-and-Artificial-Intelligence-in-higher-education-Quick-Start-guide_EN_FINAL.pdf.

University of Greenwich. (2023, March 31). Our decision to opt out of using Turnitin's AI detection tool. https://www.gre.ac.uk/articles/public-relations/our-decision-to-opt-out-of-using-turnitins-ai-detection-tool.

Weber-Wulff, D., Anohina-Naumeca, A., Bjelobaba, S., Foltýnek, T., Guerrera-Dib, J., Popoola, O., Šigut, P., & Waddington, L. (2023). Testing detection tools for AI-generated text. *International Journal for Educational Integrity, 19*, 26. https://doi.org/10.1007/s40979-023-00146-z

Whitford, E. (2022, December 9). A computer can now write your college essay – maybe better than you can. Forbes. https://www.forbes.com/sites/emmawhitford/2022/12/09/a-computer-can-now-write-your-college-essay%2D%2D-maybe-better-than-you-can/

Chapter 12
Plagiarism Without Plagiarists

Bogdan Popoveniuc

Abstract This chapter discusses the emergence of plagiarism as a significant issue in Romania, particularly following a 2012 scandal involving the former prime minister. Since then, this incident led to numerous accusations against many government officials and public persons. Notably, the decisions made by ethical committees and courts often conflicted with investigative journalism and publicly accessible evidence, as well as with each other or between similar cases. The chapter also explores how media representation influences public perception of the academic and ethical implications of plagiarism and the response of Romanian academia on this sensitive topic.

The analysis draws on a range of public sources, including media reports, court documents, and expert opinions. The second part discusses cultural and institutional factors that may explain its seemingly endemic character. The chapter concludes by suggesting some comprehensive approaches for effectively addressing plagiarism in Romania and elsewhere.

Keywords Plagiarism · Judicialisation of ethics · Meta-virtues · Integrity

Introduction

The issue of plagiarism became a matter of public interest in Romania, in 2012, when the online edition of *Nature* (Schiermeier, 2012) published the news that a significant amount of plagiarism had been found in the doctoral thesis of the then Romanian prime minister (with another former prime minister supervising his thesis). In the following years, many members of subsequent governments have faced similar accusations, to the extent that in time, at least one accused or confirmed

B. Popoveniuc (✉)
University Stefan cel Mare of Suceava, Suceava, Romania
e-mail: bpopoveniuc@usv.ro

M. Davis, C. Baptista (eds.), *Ethics and Integrity in Education (Practice)*, Ethics and Integrity in Educational Contexts 10,
https://doi.org/10.1007/978-3-031-95686-7_12

plagiarist was present in every ministerial role. The political ramifications of these plagiarism cases overshadowed what is real about the academic integrity and research practices in Romania. This can also explain why the reaction of the academic institutions was quite restrained and defensive.

Notably, the number of PhD degrees awarded saw significant growth in 2012. In that year, a total of 6259 PhD degrees were awarded. It marked the pinnacle of a nearly exponential growth rate in PhD degrees over the previous two decades. However, following this peak, the number of awarded PhD degrees declined to the levels observed in 2002, averaging around 2100 per year (I. Roşca, 2022a, 2022b).

The main characters of this saga of revealing what seems to be a widespread plagiarism practice in doctoral theses were not the universities but rather politicians, the media, and the judiciary system. This concealed the ethical nature of this practice, which was lost amidst the political rhetoric, legal considerations, and uncritical public opinion.

A large portion of the media has either remained silent on these issues or treated them as trivial news. Only a handful of journals pay close attention to this issue, with some declaring that they do not accept government funding that is disguised as information campaigns (Tuomasjukka, 2023). It is concerning that this allegation corroborates with the last report of the U.S. Department of State (2023) that exposed "excessive politicisation of media, corrupt financing mechanisms, as well as editorial policies subordinated to political parties and owners' interests". Most of the available information and findings were made public by the media and investigative journalists, but official assessments conducted by some ethical committees, prosecutors, and judges sometimes contradict what appears to be indisputable evidence.

Analysis

Political Reactions

As many public cases involved political figures who were in office at the time, it complicates the capacity to assess whether their claims of political attacks are genuine or simply a smokescreen and to discern what was really true about these allegations. Andrei Pleşu, one of the most prominent cultural figures, summarised this situation very eloquently: "*It is often said regarding Mr. Ponta's plagiarism that: «it was a manoeuvre by Băsescu» [the Romanian President at that time]. Sir, it would have been. However, the question is the following: Did he plagiarise or not? It's amazing how this real and simple issue has been turned into a complicated political matter. I hope that people judge clearly and with their minds what they see. Mr. Ponta's plagiarism is not a mere hypothesis; it is not something open to debate or analysis. It is clear evidence.*" (Mixich, 2012). This question arises in nearly all subsequent cases where plagiarism accusations emerge shortly after an individual attains a significant political position.

An analysis of media coverage reveals three primary retorts to plagiarism allegations: obtaining clearances from entities other than those in charge, initiating lawsuits or other legal proceedings to contest the claims, and promoting changes in laws or policies that clear or lessen the impact of the allegations. All three were also used in the first case with Victor Ponta's plagiarism "public trial" (Peticilă, 2020).

On the morning the verdict was to be announced, the Ministry of Education coincidentally issued an order to change the composition and procedures of CNATDCU, the national agency responsible for doctoral degree decisions CNATDCU[1] (Pantazi, 2012). In other cases, decisions made by legally competent bodies, such as universities and CNATDCU, were overturned by other government-coordinated bodies that did not have the legal authority to adjudicate these cases. Part of the accusations was ruled out apparently against the publicly available evidence. In some cases, contorted argumentation was used, even concluding in contradiction with their own analysis (e.g., CNECSDTI 40/18.05.2023, see Șercan, 2023a). The most common and convenient argument to settle the cases was on the specious grounds that the thesis was made "in accordance with the legal provisions of that time" (for Ponta's case, see CNECSDTI Decision No. 879[2]).

Other cases were litigated in court (Pușcaș, 2016). Plagiarism decisions were overturned by prosecutors (Coman, 2014; Bursa, 2013) and judges (Păvălașc, 2021), because of formal procedural defects, or administrative procedures were halted or annulled due to challenges regarding the legality of analyses conducted by universities or CNATDCU (Șercan, 2023c). It has also been suggested that ministerial structures may be complicit in assisting plagiarists by not contesting their cases or being absent during legal proceedings (Semeniuc, 2023). In some cases, the media implied potentially illicit practices even in judicial procedures (Semeniuc & Tăpălagă, 2022), but meanwhile, the cases have been investigated and closed by the judicial disciplinary board (Mihăiță, 2022; Tăpălagă & Semeniuc, 2024).

Some judicial decisions seem to contradict the principle of equality before the law. For example, one motivation relied on plagiarised excerpts from other rulings to declare the analysis of a prime minister's doctoral thesis illegal solely based on "the position occupied by him in the social hierarchy," "the professional and, specifically, the political career of the applicant," "the major impact on the credibility of the electorate and the party," and "the need for stability in the executive management of the state." The conclusion was "the fact that even the triggering, under conditions of the apparent illegality, of the verification mechanism regarding the doctoral thesis, may irreversibly affect his right to image and his professional reputation" (Bucharest Court of Appeal Decision no. 364/2022).

It is both shocking and unimaginable for a European country the compelling evidence, along with court convictions, for blackmail, death threats (Dojan, 2022), and the fabrication of compromising evidence (*kompromat*) (Dinu, 2022) targeting whistleblowers who report plagiarism. Most concerning is that the intimidation was

[1] Note: acronyms are explained in the glossary.

[2] Not publicly available on the institution's page, contrary to legal obligations.

carried out by individuals affiliated with or close to governmental structures. For months, Romania faced the risk of infringement procedures from the EU, and the second payment of 3.2 billion euros from the EU Recovery and Resilience Plans was blocked due to the flawed Romanian Whistleblowers Law, which entirely disregards evidence from anonymous reports (Pantazi, 2023).

During this period, the press highlighted at least three legislative plans intended to either stop the analysis of all past doctoral theses or to absolve them from any consequences. The Government Emergency Order 4/2016 aimed to add a new paragraph to article 168 of LEN, according to which "(4) The holder of a scientific title can request that the Ministry of National Education and Scientific Research revoke the title in question." This proposal was rejected by Romanian Constitutional Court (CCR) decision 624/2016, § 56. However, this did not prevent the High Court of Cassation and Justice (Decision No. 5321/2021) from overturning the lower court decisions 5 years later and issuing a conflicting ruling in favour of the First Deputy Attorney General's case (Hotnews.ro, 2021).

Another attempt was made to insert an amendment into an unrelated report by the Education Commission concerning the "education of underprivileged children (see Project Law P614/2021,[3] Nițulescu, 2022). If this amendment had been accepted, it would have granted complete amnesty to plagiarists with doctorates obtained before 2011. However, the opposition made the information public, and it was withdrawn. This type of initiative was not new as it had already been tried by the same Minister of Education in 2014 for clearing the former Prime Minister Victor Ponta (GEO 93/2014 M. Of. nr. 968/30 Dec. 2014, see Pantazi, 2014) but also, again, in summer 2022, in the project of the new Law of National Education. The inclusion of this provision in the new education law, which was kept secret for an extended period, was strongly backed by the same Minister of Education, who later resigned amid plagiarism accusations (Pantazi, 2022).

Meanwhile, in 2016, the CCR ruled that "the scientific title of doctor is an individual administrative act." Consequently, "the doctor's degree can be revoked by the issuing authority only during the period in which it did not enter the civil circuit and did not produce legal effects" (CCR Decision no. 624, par 39). This means that the doctoral diploma can only be revoked until it is first used. This CCR decision, along with the second decision No. 364 of 2022, removed the authority to revoke a doctoral title from both the universities and the Ministry of Education.[4] Moreover, the Court stated that "the annulment or revocation, as the case may be, of the doctoral degree/diploma cannot be carried out for these aspects [the scientific quality], but

[3] "The sanction of the withdrawal of the doctor's title provided for in para. (1) lit. b) applies to doctoral degrees developed and awarded based on the standards of quality and professional ethics established in the normative acts issued for the implementation of the National Education Law no. 1/2011".

[4] "The administrative acts that have entered the civil circuit and produced legal effects can no longer be revoked by the issuing authorities, their nullity or annulment can only be ordered by the competent court by filing an action within 1 year from the date of issuance of the act" (CCR Decision no. 624, par. 49).

only for aspects related to the legality of the conferment/attribution procedure" (CCR Decision no. 364 par 37). However, this did not stop the General Prosecutor's Office from deciding that the plagiarist cannot be accused of forgery on the ground that the doctor's diploma is only a document that certifies/attests the doctorate title and does not constitute a (false) official document in the sense of the criminal law (*De ce a fost clasat dosarul*, 2024). This interpretation also led to the annulment of the order that revoked former Prime Minister Victor Ponta's doctor title, allowing him to regain it despite clear evidence and multiple resolutions of plagiarism (Hera & Pîrvu, 2023; Edupedu.ro, 2024).

Judicialisation of Ethics

The courts' decisions and juridical considerations affect the public's understanding of plagiarism and complicate the appropriate resolution of future plagiarism cases. Reducing plagiarism to a violation of intellectual property rights disregards the fundamental values of trust, originality, and accountability that underpin scholarly work. It fosters an academic environment where researchers prioritise compliance over authentic and ethical engagement and also erodes the quality and credibility of academic research. "Granting the authority to decide on the withdrawal of a doctoral degree obtained through plagiarism to the university that organized the doctorate does not fully ensure the requirements of protecting the intellectual property rights of the author whose work was plagiarised", ruled CCR in Decision no. 624, par. 16.[5] By this, the court unfairly imposed a presumption of guilt on universities and validated the flawed argument used by plagiarists: "if it is not prohibited by law, it is permissible." Six years later, the new Higher Education Law officially granted this authority to universities, although they were required to obtain it through a court ruling.

The other issue mentioned above regarding the juridical interpretation of plagiarism was the subordination of general ethical standards to the official regulations in effect when the degree was awarded and the requirements were explicitly stipulated in administrative or legal regulations. "Concerning the meaning of this phrase ["in case of non-compliance with quality standards or professional ethics"], the Court finds that it is variable over time because the standards of quality or professional ethics vary depending on the normative regulation in force at the date of conferring/

[5] We can compare this prejudiced judicial reasoning, in contrast with those from other countries with real judicial practices and deeper understanding of society sustainability. In a similar case, the Ohio Supreme courts stated that: "To hold that a university may never withdraw a degree, effectively requires the university to continue making a false certification to the public at large of the accomplishment of persons who in fact lack the very qualifications that are certified. Such a holding would undermine public confidence in the integrity of degrees, call academic standards into question, and harm those who rely on the certification which the degree represents" (*Waliga v. Bd. of Trustees of Kent State Univ*, 22 Ohio St. 3d 55, 57 (Ohio 1986)).

attributing the doctorate" (CCR Decision no. 624, par 35). This is another common justification used by those accused of plagiarism, as well as by some within academia. While this principle can be valid in many contexts, it becomes unfounded in cases of blatant plagiarism, where there is clear intellectual theft and deception.

In my view, this is a clear case of judicial misinterpretation in which the court has also exceeded its authority. Blocking or overturning decisions made by university ethical committees through court intervention infringes on academic autonomy, which is guaranteed by the Constitution (Article 32, paragraph 6). "The fundamental aspects of university autonomy are expressed in the university charter, approved by the university senate and validated by the Ministry of Education to ensure compliance with existing legislation" (LIS, Article 11, paragraph 5). The University Charter, including the Code of Ethics, is mandatory and has legal authority within the university. From a juridical perspective, plagiarism has not a legal definition. The Romanian law recognizes only copyright infringement and plagiarism as academic misconduct cannot be categorised under it. Obligations to respect the ethics and deontology of research and development activity were not assimilated with the obligation to respect intellectual property rights in Romanian legislation as it is not under the actualized laws (Florea, 2016). In this regard, from a legal perspective, the Code of Ethics serves to supplement (*secundum legem)* what is missing (*praeter legem*) in the existing legal framework. Therefore, any court intrusion that seeks to (re)define plagiarism or judging it beyond the procedural concerns is questionable.

Academic Reactions

Caught between the Scylla of their politically dependent public funding pressure and the Charybdis of legal constraints, many universities had initially hesitated to acknowledge full responsibility openly and tried to evade a decisive resolution (PressOne, 2024; Grădinaru, 2022; Lefter, 2024; Semeniuc, 2022; Şercan, 2024b). However, in most cases, universities acknowledge their role when public scrutiny and evidence become undeniable (Bran, 2023), even though these decisions have been ruled out in courts. It is important to reiterate that the legal uncertainty surrounding universities' rights and responsibilities in handling plagiarism cases, arising from CCR and court rulings (Ghilas, 2023), has severely depressed the functioning of the academic immune system.

Measures Taken

Top-down initiatives primarily drove many corrective and pre-emptive measures. The most notable event was the closure of the Police Academy's doctoral school. After nearly 9 years, it became the first university to lose its right to award doctoral degrees due to repeated violations of ethics and academic integrity (Şercan, 2023b).

In 2018, the Ministry of Education Decree No. 3131 introduced mandatory ethics and academic integrity courses with a minimum requirement of 14 h for both master's and doctoral programs. For the undergraduate students, these courses remained optional, but universities started including it from this year. Unfortunately, the effectiveness of these new measures remains uncertain due to various factors within Romanian society. Contributing factors include a top-down decision-making structure, a centralised institutional mindset, inadequate ethical training for faculty, and the traditional resistance to critical thinking. Both teachers and students tend to disregard the importance of reflecting on personal behavior and are highly reluctant to question their own cultural practices (Popoveniuc, 2022).

In 2017, the Ministry of Education began allocating special funding aimed at improving the quality of teaching activities, with a focus on promoting adherence to ethical standards and academic integrity. It nominalises these initiatives as eligible for the annual project competition financed by the Institutional Development Fund (ME 3629, Appendix art. 6, par. G). This measure generated, at least in the first 2 years, a substantial volume of events, publications, workshops, conferences, and training sessions across the entire Romanian academic landscape. While it was a promising start, most initiatives did not go beyond writing articles and books or participating in academic discussions and events. The interest did not spread across the entire higher education system and all fields, and it remained disconnected from the general public. From my personal experience and discussions with colleagues from various fields and universities, I noticed that some of them still perceive plagiarism as less important or having a minimal impact in their field (such as in engineering).

According to Ministry of Education Order No. 5255/2021, universities were required to establish a distinct administrative structure responsible for digitising and verifying all doctoral theses completed between 1990 and 2016, as well as for training trainers in academic ethics and integrity. However, none of the universities reported any plagiarism cases up to this point, and half of them did not allocate any funding in the first years, despite some receiving millions from the government to implement these measures (Ionescu, 2024a, 2024b). The Ministry of Education neglected any measure to facilitate the analysis of doctoral works for plagiarism (Ionescu, 2023) until the enactment of the new Higher Education Law. This law diminished CNATDCU's authority to address cases of plagiarism and other academic misconduct, transferring the decision entirely to the universities (LIS, 2023).

Professor Marian Popescu, former director of the Centre for Action, Resources, Training for Academic Integrity at the University of Bucharest, dismissed the new Higher Education Law as a mere formality, a poorly crafted document produced by individuals with a superficial understanding of ethics, particularly academic ethics and deontology. He criticised its confusing amalgamation of concepts, lack of clarity, and outdated language (Popescu, 2024). The subsequent secondary legislation has created more confusion than clarity regarding the implementation of these measures. A glaring example of legislative dilettantism can be found in the *Framework Code of University Ethics and Deontology*. Among guidelines for university ethics and deontology, one finds the puzzling and illogical "principle of independence from any *moral*, *scientific*, religious, political, economic or other influence in the didactic or scientific activity" (Government Decree no. 305, Annex, art.3, par. h., *emphasis added*).

Discussion

The questions reflected in the media surrounding the notorious plagiarism cases (Semeniuc, 2024) make one wonder if there are deeper causes shaping the phenomenon of plagiarism in the Romanian public sphere. The likelihood of selection bias and disproportionate media coverage of high-profile cases suggests that we should be cautious in evaluating the extent of plagiarism in Romania. Nonetheless, the public spectacle of plagiarism cases in courts, universities, and parliament suggests the presence of systemic factors that enabled such performances.

At least three specific aspects of Romanian society can support this situation. These factors can illuminate the complexities and nuances of the problem, which may be difficult for outsiders to grasp. They also reveal the broader systemic challenges within the educational framework that would enable such practices to persist. Understanding these causes is crucial for effectively addressing the issue and promoting a culture of academic integrity.

The first factor is the culture of the political system. As the 2023 European Parliament on Academic Freedom report highlights, "the Romanian case represents a rather different political context than… other countries" (p. 187). At first glance, Romanian democracy appears to be a well-functioning system devoid of major infringements, tensions, or criticisms from civil society. However, democratic mechanisms are primarily top-down, with local communities relying on a central authority funding, while civic engagement and competence remain limited.

Undoubtedly, one of the leading causes is the poor condition of the education system. The Romanian education system has been at the bottom of the OECD (2024) ranking for years among member and candidate countries, consistently falling below average in nearly all key indicators assessed. Universities are no exception, showing similar results in international rankings - with not a single one in the top 1,000 of the 2024 Shanghai Academic Ranking. Political interference, underfunding, and restrictive regulations impede the effective management, operation, and development of universities, making it difficult for them to address ongoing challenges. Most of all, chronic political interference has a significant impact on these institutions. This fact is clear to anyone analysing the Romanian university system. "*The Romanian system struggles with the influence and abuse of political power in dealing with issues of an academic nature. In particular, recent debates have predominantly been focused on multiple cases of high-profile plagiarism, the academic community's inability to sufficiently deal with breaches of academic principles, and the influence of the political elite on elements of the new higher education law. These issues are exacerbated by a revolving door of individuals in influential positions*" (European Parliament, 2023, p. 183). In addition, universities suffer from underfunding and heavy reliance on public subsidies. They are subject to highly restrictive funding regulations that significantly constrain their hiring, compensation, and overall funding policies and practices.

The second reason is that Romanian public life is still dominated by a particular type of collectivist culture and the concentration of power (*power distance*) (David, 2015; Hofstede et al., 2010). Individual morality was not valued under Romania's communist regime. Instead, it was subsumed to collective morality and effectively suppressed. This led to a peculiar form of "tribal individualism" within the emerging democratic and capitalist settings in place today. As demonstrated in the work of Alin Gavreliuc (2011), collectivism in other cultural contexts places a strong emphasis on the significance of communal solidarity. Romanian collectivism can be characterised as the instrumental use of the group to pursue individual objectives. True collectivism prioritises the group, while the Romanian version exploits it for individual benefit in an unfair social setting, compromising the social game. The moral disengagement and low moral responsibility for one's deeds is explained by the dominance of conservative values of security, (orthodox) religiousness, tradition, family, and obedience, together with their fundamental attitudes such as fatalism, "assistanceness", public disengagement, and behavioural opportunism. As one astute internet meme aptly captures, post-revolutionary Romanian capitalist democracy is defined by caviar socialism and crony capitalism, alongside the latest trend of TikTok Christianity. As Alin Gavreliuc argues, "*what is disturbing to us is not the fact that there are deviant behaviors, but their extent in the behavioral ensemble, the naturalness of the mutual stealth that governs the social game, the lack of guilt, shame, responsibility, from the belief that if you have the opportunity to deceive the others to win, it is the reward for the times when you were scammed in your turn.*" (A. Roşca, 2022a, 2022b; see also Gavreliuc, 2011). The social pressure of this micro-morality compels individuals to act with utmost consideration towards family, friends, and close acquaintances, far exceeding the expectation of similar behaviour toward others. This moral mentality starkly contrasts broader societal ethics, which emphasise universal obligations and fairness. Instead, moral responsibilities are deeply rooted in personal relationships, often leading individuals to privilege the needs of family and friends over the collective good - to the point where unlawful actions are readily perceived as morally legitimate, provided they serve one's immediate circle. Certainly, there is nothing wrong with prioritising close ones; in fact, it underpins sustainable human communities and is a fundamental source of moral consciousness. However, it becomes problematic when it flagrantly violates the rights of others and conspicuously ignores their needs. In such a morally conflicting culture, individuals face challenges in balancing their personal loyalties with their responsibilities to society as a whole.

This leads to the strange belief that anyone who takes an active legal stance toward plagiarism is suspected of bad intentions and immorality and even blamed. Plagiarism is wrong, but plagiarists are your colleagues, people, and professionals like you, who have the right to make mistakes. Exposing or judging them is wrong. This is the Romanian Balkan mentality. It may sound dystopian, but the decade-long plagiarism saga outlined here shows that Romanian society still struggles to address and sanction unethical behaviour. This, I believe, reflects the broader Romanian society's inability to publicly confront and accept that the communist past distorted our moral compass (Hotnews.ro, 2023).

The third factor contributing to the current state of academic ethics is that students often do not feel a sense of belonging within the academic community, and neither do some staff members. Students experience school as an alien environment they must pass through, while staff members refer to it as an ordinary job that, despite their social prestige and responsability, primarily serves as a means to earn a living. There is a noticeable lack of awareness regarding their vocational and social expectations placed upon them. Universities are predominantly perceived as providers of credentials required for professional qualification. Consequently, many students assume that access - whether paid or free - entails an automatic entitlement to graduation, irrespective of academic engagement or performance. This environment often prioritises short-term gains over long-term educational integrity, creating a landscape where ethical considerations are sidelined, and accountability is lacking.

During my academic ethics and integrity courses at the master level, I ask students to remember how they felt when they cheated or helped others to cheat on an exam. Very few deny they did this, and even fewer see it as a moral dilemma. Violating the principles of academic fairness is primarily assessed based on its potential negative consequences rather than being viewed as a moral transgression. The refusal to unethically assist colleagues is driven not by a recognition of the act's wrongdoing, but rather by selfish motives such as protecting one's own interests (e.g. scholarships), epicaricacy (*schadenfreude*), egoism, or indifference. They are aware they are breaking a rule, but not that it is deeply unethical. These empirical facts should inform our search for a solution, rather than relying on theoretical projections of what ought to be done.

As educators, we bear a significant portion of the responsibility. Professions such as teachers, clergy, magistrates, doctors, or police officers, among others, are characterised by their commitment to public service, strict ethical standards, specialised training, heightened accountability, direct impact on people's lives, and reliance on community trust. These attributes set them apart from most other professions. One must try harder to fulfil one's social role to equate one's social status. They play a crucial role in supporting the backbone of any healthy society, meaning education and morality, justice, health, or security. When individuals in these roles lack personal development and an appropriate understanding of their responsibilities, it can severely hinder the effective functioning of society. As educators, we must recognise that our role extends beyond merely imparting specialised knowledge, training skills, and particular competencies. We also have a profound responsibility to instil moral values and integrity in our students. Fostering a solid ethical foundation is essential for their personal and professional development, as well. By emphasising the importance of honesty, accountability, and respect, we prepare our students to navigate the world's complexities with a sense of purpose and responsibility. Ultimately, cultivating these values alongside academic knowledge ensures that our students become not only competent professionals, but also civically and ethically engaged individuals. If we do not believe in our mission, we cannot expect our students to believe in these values.

Recommendations

Romanian universities struggle to fulfil their public mission within a challenging social and cultural context. They do not receive the same level of respect and significance in society as their Western counterparts. On the path toward healing the academic setting from past plagiarism wounds and strengthening its immunity system, the effort must come not only from the academic community, but also from the state and all other stakeholders. There are several dimensions that need to be addressed or strengthened.

First, there is an urgent need for *clarification and improvement of the legislative framework*. As we already seen, significant progress has been made at the institutional and legislative level regarding the issue of plagiarism. The new Higher Education Law (LIS) and the Code of doctoral studies (Ministry of Education Order no. 305/2024) passed on higher education institutions the responsibility in reporting and solving cases of plagiarism, depicts in detail and with rigour the verification and validation procedures and increased to 3 months the period in which the thesis is in public transparency before the official defence. The predictable effect is to stimulate preventive measures on the part of universities.

It remains to be seen whether the mechanisms outlined for resolving potential appeals will function effectively or if obstacles will arise. Currently, the institutional role of CNATDCU in this process has become unclear. Its responsibilities regarding doctoral theses have been diluted, and it has lost significant autonomy, as it is now subordinated to the Ministry of Education. Strengthening the role of university ethics commissions in matters related to plagiarism, should also mean respecting their decisions, without being annulled by the courts that should resume to fulfil the legal confirmation of the withdrawal of the title. The ÎCCJ should ensure consistency and uniformity in the application of the law, avoiding subjective interpretations and contradictory decisions by the courts. Ethics committees must have the power to make recommendations and impose sanctions without fear of judicial review. Therefore, it would be beneficial to have a clear definition of plagiarism, and its consequences outlined in legislation. Additionally, it must be established that the academic environment has an exclusive role in making decisions grounded in academic expertise. This will ensure that their decisions are respected, and the internal accountability mechanism will be effective. LIS also instantiates ethics and academic integrity courses, at their rightful place, at the bachelor's level in addition to the master's and doctorate. However, it only specifies that the educational plans at the three levels "contain disciplines of ethics and academic integrity, including research ethics" (LIS, article 28 paragraph 10) without stating that these subjects are mandatory. Only at the doctoral level is it stated that these are "obligatory to be completed by all enrolled doctoral students" (Ministry of Education Order no. 305/2024, article 12, paragraph 20).

Secondly, measures have already been taken to *strengthen monitoring and prevention mechanisms*. The State already provided universities with grants or direct funding for developing the capacity of institutions to systematically analyse (past) doctoral theses for the identification of plagiarism and investing in digital technologies and tools for its automated detection in doctoral theses, master and bachelor's dissertation. Universities should use these funds to finance training and specialisation for teaching and research staff in the areas of ethics and academic integrity. The effectiveness of these measures will depend on the seriousness with which they are implemented, rather than merely being a formality on paper. It stands to reason that external peer-review and international cooperation have beneficial effects on preventing plagiarism and academic misconduct. An important aspect that has already been implemented is the mandatory inclusion of an external member on the supervision committee, now known as the supervision and academic integrity committee. Some universities already are recommending a higher standard by including a member from a foreign university, at least for the thesis public defence committee. In this way, better external monitoring would be ensured. Some of the key indicators used by the evaluation agency ARACIS to assess doctoral schools are already focused on enhancing this aspect. Specifically, indicator B.3.2.1 and B.3.2.2 stipulate that the presence of an external evaluator in the public defence commissions must not exceed two theses coordinated by the same PhD supervisor within 1 year. Indicator B.3.2.2 requires that over the entire five-year period, one external member should not serve on more than 30% of the commissions in the respective field. Nevertheless, questionable conduct can be noticed among university consortia that informally recommend or even require the exclusion of referees from Romanian universities that are not part of the partnership. In my view, this measure undermines the primary selection criterion of academic expertise that is essential for maintaining proper academic standards.

At the same time, exchanging good practices and learning from the experiences of other countries in combating plagiarism is essential. Engaging in international networks focused on academic integrity will allow institutions to benefit from their expertise and resources. Additionally, organising events, workshops, and awareness campaigns will help promote these values among students and faculty.

Another essential measure would be a comprehensive *review of the academic rewards and promotion system*. At the university level, it is important to eliminate distorted forms of merit-based salary incentives that encourage quantitative publication at the expense of quality and integrity. Instead, a broader range of other contributions to responsible research and academic activity should be integrated, such as "peer review for grants and publications, mentoring, outreach and knowledge sharing", as recommended by the 5th Hong Kong Principle for assessing researchers (Moher et al., 2020). Additionally, it is essential to reduce the excessive teaching and administrative responsibilities of PhD coordinators to ensure they have adequate time to closely monitor the progress of doctoral students.

Also, it is important *to eliminate the temptations associated with pursuing a PhD from outside academia*, by removing the inadequate incentives that exist in other professional sectors. Several professional advantages motivate individuals to obtain doctoral titles for reasons unrelated to a genuine interest in research or an academic career. These incentives include access to the bar association, eligibility for public expert positions, entry into the judiciary system, and advancements in military, police, clergy, or other careers circumventing longer periods of specialised training. Additionally, individuals may gain material benefits such as salary increases or pension enhancements, even though research skills are not relevant to the quality or specifics of these fields.

Not the least, there are necessary efficient measures for *consolidation of university autonomy*. Ensuring the independence of universities from political interference and funding pressures in academic ethics decision-making is essential. This can be achieved by imposing certain restrictions on university management members, such as prohibiting them from being part of the governing bodies of political parties or holding politically appointed public positions. At university level, measures should be taken to democratise decision-making and to reduce the near-total control of the Rectors. Allowing faculty members to elect deans, rather than having them appointed by the Rector as was the case before the 2011 education reform, would promote a more democratic and decentralised mechanism of university management. Decentralising the national academic system by reducing its dependence on political decisions is crucial. This can be implemented by ensuring more independent and diversified funding so that universities become less dependent on political decisions for their budgets. Here, one could list legislative initiatives that include reducing or conditioning budget allocations based on more qualitative and efficiency criteria, measures to encourage the involvement of private sector universities through public-private partnerships, the creation of investment funds dedicated to education, legislative facilities for donations from alumni and non-governmental organisations or scholarships funded by sponsors, promoting foreign language study programs and offering discounts or scholarships for international students, improving the existing model of (additional) performance-based funding, supporting innovation and entrepreneurship through business incubators to generate startup income and spin-offs, capitalising the patents, facilitating the attraction of grants and European funds for research projects, offering consulting services, professionalisation and conversion courses for companies, etc.

In my opinion, the academic community has learned its lesson regarding copy-paste plagiarism. There will likely be fewer cases overtime. However, we must be realistic and acknowledge that an honest confrontation with the past is not feasible in a society deeply traumatised, civically weakened, and morally atrophied by decades of dictatorship. The spectres of past academic misconduct will continue to haunt for some time, and their exorcism will be postponed until they no longer affect those involved, much like the protracted legal proceedings concerning the Romanian revolution, which have been restarted after 30 years of delays (Hotnews.ro, 2023).

The significant systemic measures implemented to prevent and address academic misconduct should be acknowledged. However, for these solutions to be effective, they must be applied consistently and continuously, engaging both authorities and the academic community. The danger lies in the previously mentioned cultural weaknesses, which can lead to superficial implementation of measures, a lack of awareness about the importance of promoting integrity, peer pressure to overlook 'minor' violations of proper academic conduct, inadequate supervision, the disregard of the importance of ethics and academic integrity courses, and the insufficient training of the teachers who deliver these courses.

Unfortunately, a new and more insidious danger looms on the horizon. In the absence of a development of ethical organisational culture in the university environment, subtler forms of plagiarism, such as the appropriation of ideas, lack of originality, and redundant research works, may emerge. Also, the potential use of technologies based on large language models (LLMs) increase the efforts for prevention and identification, requiring more expertise and resources. The ability to address these new challenges comes from fostering integrity within the university environment. Reducing academic and research imposture necessitates a more nuanced and comprehensive approach to academic misconduct. This can be realised by promoting an organisational ethical culture that prioritises different values than those found in utilitarian or deontological frameworks, which often lead to negative incentives for questionable conduct or conformity. Consequently, universities should integrate a virtue ethics approach in their teaching methods to cultivate the ethical conduct in their students. Under these challenging conditions, the most effective approach is to strengthen students' sense of belonging to the academic institution, fostering an attachment resembling, because similar it cannot be, to that which they feel toward family, church, friendships, and other socialising structures. This means not as an allegiance to a university as a formal institution, but rather as a virtual community of authentic thinkers and valid knowledge producers.

Conclusion

Alin Fumurescu expressively summarises the problem of plagiarism in Romania. He illustrates the state of the plagiarism issue through two artful paraphrases of two significant figures in Romanian culture: the national poet Mihai Eminescu and the playwright Eugen Ionescu. "Plagiarism was while it went unnoticed/ Today we see it, and it is not", and fighting against plagiarism is like art, "absolutely useless, yet this uselessness is absolutely necessary" (in Şercan, 2024a).

Nevertheless, plagiarism is only the tip of the iceberg. As a transgression of appropriate conduct in research and academia, it is merely primitive, rude, and deeply dishonouring behaviour. False feathers for small swindlers who cheat their fellows to gain undeserved credit and benefits. However, it is benign compared with other types of misconduct, such as fabrication or falsification of data, which discredits scientific knowledge and not only its public image but also alters the very drive of scientific progress and puts the sustainable development of society in peril. Plagiarism only expresses the moral indignity of its authors - their deep disdain for education, research, and culture. However, the real issues are not individual cases. We should focus on the systemic aspects of academic cultures. The commodification of knowledge creates a framework that incentivises dishonesty including a "publish or perish" culture that prioritises quantity over quality, grant chasing, metrics-driven evaluation, and publication prestige bias. Additionally, we must consider the social context - cultural mentality, habits, customs, and political environment - in which the university operates, as well as the underlying psychological contributing to cheating if we aim to preserve and promote the university's mission. This is a most pressing task in this new "post plagiarism" epoch when the evident and rude classical concept of plagiarism is being disrupted by the rise of advanced technologies like AI, requiring rethinking the notions of authorship, attribution, and academic integrity (Eaton, 2023).

Sadly, despite all the events and legislation changes described in this chapter, there has been no significant shift in public perception in Romania regarding plagiarism as an ethical transgression, along with other questionable research practices. The core ethical and cultural implications of plagiarism were obscured by misleading debates on its political weaponisation and the less significant, even debatably, legal ramifications of copyright infringement. Furthermore, it is disheartening that university leadership, public authorities, and even some members of the academic staff tacitly yet strongly discourage any overt actions - such as official complaints or whistleblowing - beyond mere rhetorical support or unconvincing conduct addressing plagiarism and other forms of academic misconduct. The authorities' strategy on these issues appears to be contradictory. An illustrative example is the authorities conduct following recent cases of sexual harassment. The government announced legislative changes that would permit anonymous reporting, but in the meantime, the Ministry of Education issued the Framework Regulation on the Organization and Operation of University Ethics Committees, which explicitly states that anonymous notifications and complaints are not considered" (Article 6, paragraph 3). In the existing Framework Code of Ethics in Universities such reports were only implicitly prohibited through its referral guidelines (Ionescu, 2024a, 2024b). Although the legal framework allows ethics commissions to self-report, the elimination of anonymous reporting is discouraging.

The portrayal of public scandals in the media, along with the image strategies employed by public figures – allegedly aided by judicial and state institutions to clean up their reputations and evade legal consequences - has a detrimental impact on public perception. The students fail to understand the seriousness of plagiarism due to the excessive politicisation of the subject and the lack of attention paid to academic integrity (Voiculescu, 2022). In universities, the atmosphere looks more like one of anxious anticipation than indignation. Are plagiarism skeletons lurking in their closets, too? They seem more concerned with how to counteract the public perceptions of their image if something comes up than confronting possible past mistakes and proactively building a culture of integrity.

At the same time, passing the decision onto court arbitration is definitely inappropriate. Marian Popescu served as the president of the ethics committee that issued the plagiarism verdicts for Victor Ponta, the then-prime minister, and later for Gabriel Oprea, the then-acting deputy prime minister. In his opinion, questionable motivations of ethical committees and courts arise from a "brutal manipulation of the legislative framework they invoked" (in Şercan, 2024b). The current judicialization of academic ethics is detrimental to the organisational culture of universities, much like the broader phenomenon of judicialization in politics. The former undermines the fundamental pillar of the academic system, integrity, disrupting the genuine relationships between ethical standards and scholarly accountability. The latter undermines the fundamental pillars of the modern state and democracy, disrupting the natural relationship between morality and justice. A society is sustainable only if its foundations of justice and legal systems are grounded in moral principles and ethical considerations rather than the other way around. Judicialization of academic ethics illustrates a more severe form of Campbell's law, highlighting how excessive reliance on quantitative or formalised social indicators for decision-making can lead to adverse outcomes. This is particularly true when the indicator does not accurately reflect the nature of the social dimension it is meant to measure, such as the legal aspects of ethical conduct. The more formal legal criteria are employed to assess the social acceptability of an action, the higher the risk of corruption and distortion in the very social processes these criteria are intended to evaluate. In the case of plagiarism, copyright is the last concern. When inappropriate judicial evaluations obscure the fundamental ethical implications of plagiarism, it reduces the moral outrage associated with this practice. Consequently, the likelihood of eliminating misconduct in the academic climate and promoting a culture of integrity in research is significantly reduced. It is worth noticing that this law-ruling copyright enforcement led to a questionable situation that hinders scientific progress at the global level. The voluntary contributions of millions of researchers are often hidden behind unaffordable corporate paywalls, separating valuable knowledge from both the public and their peers.

The problem of plagiarism in Romania extends far beyond individual cases covered in this chapter. It reflects a more profound systemic failure within the academic culture. The persistent public scandals and the seeming reluctance of university leadership to confront these issues contribute to a culture of wilful moral blindness and inaction, where public relations exercises overshadow the true implications of plagiarism. Unless we address the underlying motivations that foster dishonesty and create an environment that prioritises ethical conduct, academic integrity will continue to be compromised. Ultimately, a sustainable society relies on a solid educational framework rooted in moral principles. Without this foundation, the cycle of misconduct will continue, hindering the genuine progress of civilization and eroding the essential pillars of knowledge and trust in the University, which is one of the fundamental institutions of democracy.

Glossary of Acronyms for National Legal Bodies Regulating Higher Education Institutions

ARACIS (Agenția Română de Asigurare a Calității în Învățământul Superior) [Romanian Agency for Quality Assurance in Higher Education]. "The mission of ARACIS is to conduct external evaluations to assess the quality of education provided by higher education institutions and other organisations that offer specialised higher education programs." This evaluation process is designed to ensure the accreditation of these programs. Web page: https://www.aracis.ro/.

CCR (Curtea Constituțională a României) [Constitutional Court of Romania]. Web page: https://www.ccr.ro/en/5864-2/.

CEMU (Consiliului de Etică şi Management Universitar) [The University Ethics and Management Council] is an advisory body of the Ministry of National Education, without legal personality. The mission of the Council of Ethics and University Management is to develop a culture of ethics and integrity in Romanian universities and to determine and support universities to develop and put into practice, in an organised, transparent, and efficient manner, ethics and university integrity policies, in accordance with the legal provisions. Web page: http://www.cemu.ro/.

CNATDCU (Comisia Națională de Atestare a Titlurilor, Diplomelor si Certificatelor Universitare) [The National Council for Attestation of University Titles, Diplomas and Certificates] is an advisory body, without legal personality, of the Ministry of National Education. It mainly, among other attributions, evaluates the thesis and dossier for awarding the doctoral degree, analyses and decides on reports of non-compliance with professional ethics standards, and examines the existence of plagiarism. (Rules governing the functioning and organisation of CNATDCU (Ministry Order Nr. 4621/2020 23 June, M.Of. nr. 564 /2020 June 29). Web page: http://www.cnatdcu.ro/.

CNFIS (Consiliului Național pentru Finanțarea Învățământului Superior) [National Council for the Financing of Higher Education] is a national advisory body of the Ministry of Education, develops principles and methods for distributing public funds to state universities in Romania. Through its activity, CNFIS promotes the continuous increase of quality in the higher education system in Romania, ensuring that all citizens have equal training opportunities in higher education. Web page: http://www.cnfis.ro/finantare/dezvoltare institutionala/#.

CNECSDTI (Consiliul Național de Etică a Cercetării Ştiințifice, Dezvoltării Tehnologice şi Inovării) [National Ethics Council of Scientific Research, Technological Development and Innovation]. https://cnecsdti.research.gov.ro/.

CNR (Consiliul Național al Rectorilor) [The National Council of Rectors] is a consultative, non-governmental, politically independent, and non-profit national body, established under OG 26/2000 and the National Education Law no. 1/2011." Web page: https://devtv.uvt.ro/.

DNA (Direcția Națională Anticorupție) [National Anticorruption Directorate] "The National Anticorruption Directorate (DNA) is a prosecutor's office specialised in combating high and medium-level corruption. It was set up as a necessary tool for discovering, investigating, and indicting high and medium-level corruption cases.

As a result of its activity, DNA contributes to reducing corruption, giving its support for a democratic society close to the European values." Web page: https://www.pna.ro/index.xhtml.

ÎCCJ (Înalta Curte de Casație și Justiție) [High Court of Cassation and Justice] the highest court in Romania's judicial system, which „ensures the uniform interpretation and application of the law by the other courts, according to its competence."

Documents from Legislative Bodies, Courts, and Other Normative Authorities

Bucharest Court of Appeal Decision no. 364/2022 March 01. https://www.rejust.ro/juris/59235e638.

CNECSDTI Decision No. 41, 2023 May 18. https://cnecsdti.mcid.gov.ro/wp-content/uploads/2023/Hot%c4%83r%c3%a2rea%20CNECSDTI%20nr.%2040%20din%2018.05.2023%20%c8%99i%20Raportul%20final%20nr.%2040.pdf.

CNECSDTI Decision No. 879, 2012 July 18. https://cdn.edupedu.ro/wp-content/uploads/2020/07/document-2012-07-19-12835202-0-hotarearea-878-raport-final-879-1.pdf) https://cnecsdti.research.gov.ro/hotarari/hotarari-in-perioada-2012-2015/.

Romanian Constitution. https://www.cdep.ro/pls/dic/site.page?id=339.

Constitutional Court Decision 364/2022 June 6. (Published in Monitorul Oficial (M.Of.) [Official Gazette], Part 1, nr. 831/2022 August 24). https://www.ccr.ro/wp-content/uploads/2022/08/Decizie_364_2022.pdf.

Constitutional Court Decision 624/2016 October 26 (M.Of. Part 1, nr. 937 /2016 November 22). https://www.ccr.ro/wp-content/uploads/2020/07/Decizie_624_2016.pdf.

European Parliament (2023). *EP Academic Freedom Monitor 2023*. https://www.europarl.europa.eu/RegData/etudes/STUD/2024/757798/EPRS_STU(2024)757798_EN.pdf.

Government Decree no. 305/2024 [for approving the Framework Code of University Ethics and Deontology]. (M.Of Part 1, nr. 327/2024 April). https://legislatie.just.ro/Public/DetaliiDocumentAfis/281569.

Government Emergency Order 93/2014 (M. Of. nr. 968/2014 December 30). https://monitoruloficial.ro/Monitorul-Oficial%2D%2DPI%2D%2D968%2D%2D2014.html.

Government Emergency Order 4/2016. (M.Of. Part 1, nr. 182/2016 March 10). https://www.cdep.ro/pls/proiecte/upl_pck2015.proiect?idp=15544.

High Court of Cassation and Justice Decision no. 5321/2021 November 4. https://www.scj.ro/1093/Detalii-jurisprudenta?customQuery%5B0%5D.Key=id&customQuery%5B0%5D.Value=189562#highlight=##.

LEN (Legea Educației Naționale) [National Education Law]. (no. 1 of January 5, 2011, M.Of. Part 1, nr. 18/2011 January 10). https://legislatie.just.ro/Public/DetaliiDocument/125150.

LIS (Legea învățământului superior) [Higher Education Law]. (no 1992023 July 4, M.Of. Part 1, no. 614/2023 July 5). https://legislatie.just.ro/Public/DetaliiDocument/271898.

Project Law P614/2021. https://www.cdep.ro/pls/proiecte/upl_pck2015.proiect?idp=18405.

The Order of the Minister of Education no. 3629/2017. (M.Of. Part 1, no. 278/2017 April 20) [regarding the approval of the Methodology for the allocation and use of the fund for the institutional development of state higher education institutions]. https://legislatie.just.ro/Public/DetaliiDocument/188456.

The Order of the Minister of Education no. 3131/2018. (M.Of. Part 1, no. 140/2018 February 14) [regarding the inclusion in the education plans for all university study programs organised in the higher education institutions of the national education system, of ethics and academic integrity courses]. https://legislatie.just.ro/Public/DetaliiDocument/197844.

The Order of the Minister of Education No. 5255/2021. (M.Of. Part 1, no. 888/2021 September 16) [regarding the verification of compliance with university ethics and deontology in the preparation of doctoral theses from the period 1990–2016.]. https://legislatie.just.ro/Public/DetaliiDocument/246418.

The Order of the Minister of Education No. 3020/2024. (M.Of. Part 1, no. 56/2021 January 22) [for the approval of the Framework Regulation on doctoral university studies]. https://www.edu.ro/sites/default/files/fisiere%20articole/OM_3020-2024.pdf.

The Order of the Minister of Education No. 6.869/2024. (M.Of. Part 1, no. 990/2024 October 3) [for the approval of the framework Regulation on the organization and operation of university ethics commissions]. https://legislatie.just.ro/Public/DetaliiDocument/289094.

U.S. Department of State. (2023). 2023 Country Reports on Human Rights Practices: Romania. https://www.state.gov/reports/2023-country-reports-on-human-rights-practices/romania/.

Waliga v. Bd. of Trustees of Kent State Univ, 22 Ohio St. 3d 55, 57 (Ohio 1986). https://casetext.com/case/waliga-v-bd-of-trustees-of-kent-state-univ.

References

Bran, M. (2023, January 20). En Roumanie, le gouvernement fragilisé par une série d'affaires de plagiat [In Romania, the government weakened by a series of plagiarism cases]. Le Monde.fr. https://www.lemonde.fr/international/article/2023/01/20/en-roumanie-le-gouvernement-fragilise-par-une-serie-d-affaires-de-plagiat_6158648_3210.html.

Bursa. (2013, May 22). Parchetul susţine că Victor Ponta nu a plagiat [The prosecutor's office claims that Victor Ponta did not plagiarize]. Bursa. https://www.bursa.ro/parchetul-sustine-ca-victor-ponta-nu-a-plagiat-29546026.

Coman, D. (2014, March 17). Victor Ponta's plagiarism file: The court will decide on Monday whether the investigation will be resumed [Victor Ponta's Plagiarism Case: Court to Decide on Investigation Resumption on Monday]. *Digi24*. https://www.digi24.ro/stiri/actualitate/justitie/dosarul-de-plagiat-al-lui-victor-ponta-instanta-decide-luni-daca-ancheta-va-fi-reluata-215299.

Ionescu, C. (2024a, October 4). Sesizările anonime în universităţi—blocate explicit prin noul Regulament-cadru de organizare şi funcţionare a comisiilor de etică, în absenţa schimbărilor legislative promise de Guvernul Ciolacu după cazurile de abuzuri sexuale [Anonymous complaints in universities - explicitly blocked by the new framework Regulation on the organization and operation of ethics commissions, in the absence of the legislative changes promised by the Ciolacu Government after the cases of sexual abuse]. *Edupedu.ro*. https://www.edupedu.ro/sesizarile-anonime-in-universitati-blocate-explicit-prin-noul-regulament-cadru-de-organizare-si-functionare-a-comisiilor-de-etica-in-absenta-schimbarilor-legislative-promise-de-guvern-noile-regul/.

David, D. (2015). Psihologia poporului român. Profilul psihologic al românilor într-o monografie cognitiv-experimentală [The psychology of the Romanian people. The psychological profile of Romanians in a cognitive-experimental monograph]. Polirom.

De ce a fost clasat dosarul privind plagiatul lui Nicolae Ciucă. Parchetul General: Fapta s-a prescris în anul 2008. (2024, May 6). *Euronews.ro*. https://www.euronews.ro/articole/de-ce-a-fost-clasat-dosarul-privind-plagiatul-lui-nicolae-ciuca-parchetul-general

Dinu, C. (2022, August 4). Culisele operațiunii kompromat. [Behind the scenes of the kompromat operation]. *Hotnews.ro*. https://www.hotnews.ro/stiri-esential-25486916-culisele-operatiunii-kompromat-emilia-sercan-avut-soc-duci-politie-reclami-politia-vinde-mafiotilor.htm.

Dojan, S. (2022, July 21). Condamnare cu suspendare după amenințarea cu moartea a unui jurnalist. [Conviction with suspendation after death threat to a journalist]. *Europa Liberă România*. https://romania.europalibera.org/a/emilia-sercan-plagiat-condamnare-iccj/31953939.html.

Eaton, S. E. (2023). Postplagiarism: Transdisciplinary ethics and integrity in the age of artificial intelligence and neurotechnology. *International Journal for Educational Integrity, 19*, 23. https://doi.org/10.1007/s40979-023-00144-1

Edupedu.ro. (2024, September 20). Ordinul prin care i s-a retras titlul lui doctor fostului premier Victor Ponta a fost anulat de Înalta Curte de Casație și Justiție, în pofida unei hotărâri definitive din 2020. [The High Court of Cassation and Justice annulled the order that revoked the title of doctor from former Prime Minister Victor Ponta, despite a final decision made in 2020]. *Edupedu.ro*. https://www.edupedu.ro/ordinul-prin-care-i-s-a-retras-titlul-lui-doctor-fostului-premier-victor-ponta-a-fost-anulat-de-inalta-curte-de-casatie-si-justitie-in-pofida-unei-hotarari-definitive-din-2020/.

Florea, S. (2016, September 19). Plagiatul și încălcarea drepturilor de autor.[Plagiarism and copyright infringement]. *Juridice.ro*. https://www.juridice.ro/467536/plagiatul-si-incalcarea-drepturilor-de-autor.html.

Gavreliuc, A. (2011). Româniile din România. Individualism autarhic, tipare valorice transgeneraționale și autism social [Romanias from Romania. Autarchic individualism, transgenerational value patterns and social autism]. Editura Universității de Vest.

Ghilas, O. (2023). UBB nu îl dă în judecată pe Lucian Bode, pentru retragerea titlului de doctor]. Monitorulcj.Ro. http://www.monitorulcj.ro/educatie/106311-ubb-nu-il-da-in-judecata-pe-lucian-bode-pentru-retragerea-titlului-de-doctor-nu-sunt-argumente-pentru-o-astfel-de-sustinere.

Grădinaru, A. (2022, November 1). UBB admite că nu a verificat toată teza de doctorat a lui Bode ci "doar ce s-a reclamat". [UBB admits that it did not check all Bode's doctoral thesis but "only what was claimed"]. *Europa FM*. https://www.europafm.ro/ubb-admite-ca-nu-a-verificat-toata-teza-de-doctorat-a-lui-bode-ci-doar-ce-s-a-reclamat-nu-exista-capacitate-pentru-asa-ceva-spune-reprezantantul-comsiei-de-etica/.

Hera, M., & Pîrvu, L. (2023, February 22). Victor Ponta folosește decizia Curții Constituționale privind plagiatul pentru a-și recâștiga titlul de doctor [Victor Ponta uses the Constitutional Court's decision on plagiarism to regain his doctorate] *Hotnews.ro*. https://www.hotnews.ro/stiri-esential-26100474-victor-ponta-foloseste-decizia-curtii-constitutionale-privind-plagiatul-pentru-recastiga-titlul-doctor-comis-ilegalitate-cronica-celui-mai-mare-scandal-plagiat-din-romania.htm.

Hofstede, G., Hofstede, G. J., & Minkov, M. (2010). *Cultures and organizations: Software of the mind*. 3rd ed.

HotNews.ro. (2021, November 5). Prim-adjunctul procurorului general Bogdan Licu, câștigă în instanță dreptul de a renunța benevol la titlul de doctor obținut cu o teză suspectată de plagiat.

[The first deputy of the general prosecutor, Bogdan Licu, wins in court the right to willingly renounce the doctorate obtained with a thesis suspected of plagiarism]. *HotNews.ro*. https://hotnews.ro/prim-adjunctul-procurorului-general-bogdan-licu-cstiga-n-instanta-dreptul-de-a-renunta-benevol-la-titlul-de-doctor-obtinut-cu-o-teza-suspectata-de-plagiat-163533.

Hotnews.ro. (2023, February 24). Dosarul Revoluției, poveste fără sfârșit. [The Revolution file, a never-ending story]. *Hotnews.ro* https://hotnews.ro/dosarul-revolutiei-poveste-fara-sfrsit-instanta-suprema-l-trimite-la-curtea-de-apel-bucuresti-invocnd-ca-nu-are-competenta-pentru-ca-atunci-ion-iliescu-nu-era-presedinte-79639.

Ionescu, C. (2023, February 23). Nicio măsură care să ducă la deblocarea analizei doctoratelor plagiate nu este propusă pentru 2023 de Ministerul Educației [No measure leading to unblocking the analysis of plagiarized doctorates is proposed for 2023 by the Ministry of Education]. *Edupedu.ro*. https://www.edupedu.ro/nicio-masura-care-sa-duca-la-deblocarea-analizei-doctoratelor-plagiate-nu-este-propusa-pentru-2023-de-ministerul-educatiei-in-cel-mai-nou-raport-privind-starea-invatamantului-superior-in-noiembrie/.

Ionescu, C. (2024b, February 14). Banii dați fiecărei universități de stat pentru verificarea doctoratelor scrise între 1990-2016 [The money given to each state university for checking doctorates written between 1990-2016]. Edupedu.ro. https://www.edupedu.ro/banii-dati-fiecarei-universitati-de-stat-pentru-verificarea-doctoratelor-scrise-intre-1990-2016-jumatate-nu-au-cheltuit-in-2021-nimic-desi-unele-au-primit-milioane-de-lei-in-doi-ani-nici una-nu-a/.

Lefter, Ș. (2024, August 20). Două tabele și 43 de rânduri fără citare—Plagiatul lui Mircea Geoană în teza de doctorat, stabilit de Comisia de etică a ASE București [Two tables and 43 lines without citation—Mircea Geoană's plagiarism in his doctoral thesis, established by the Ethics Commission of ASE Bucharest]. *Edupedu.ro*. https://www.edupedu.ro/doua-tabele-si-43-de-randuri-fara-citare-plagiatul-lui-mircea-geoana-in-teza-de-doctorat-stabilit-de-comisia-de-etica-a-ase-bucuresti-document-avand-in-vedere-procentul-redus-de -plagia/.

LIS (2023). (Legea învățământului superior) [Higher Education Law]. (no 1992023 July 4, M.Of. Part 1, no. 614/2023 July 5). https://legislatie.just.ro/Public/DetaliiDocument/271898

Mihăiță, F. (2022). Parchetul general investighează modul în care a fost repartizat dosarul privind plagiatul lui Ciucă. [The General Prosecutor's Office is investigating how the file regarding Ciucă's plagiarism was distributed.]. *Europa FM*. https://www.europafm.ro/parchetul-general-investigheaza-modul-in-care-a-fost-repartizat-dosarul-privind-plagiatul-lui-ciuca-judecatorul-s-a-pensionat-dupa-decizie/.

Mixich, V. (2012, July 12). Andrei Plesu: Eu as defila prin tot Parlamentul Romaniei, printre toti guvernantii Romaniei, si i-as intreba: "Ce stii sa faci?" [Andrei Plesu: I would parade through the entire Romanian Parliament, among all the governors of Romania, and I would ask them: "What do you know to do?"]. *Hotnews*. https://www.hotnews.ro/stiri-esential-12887975-video-andrei-plesu-defila-prin-tot-parlamentul-romaniei-printre-toti-guvernantii-romaniei-intreba-stii-faci.htm.

Moher, D., Bouter, L., Kleinert, S., Glasziou, P., Sham, M. H., Barbour, V., Coriat, A.-M., Foeger, N., & Dirnagl, U. (2020). The Hong Kong principles for assessing researchers: Fostering research integrity. *PLoS Biology, 18*(7), e3000737. https://doi.org/10.1371/journal.pbio.3000737

Nițulescu, L. (2022, March 7). Amnistie pentru plagiatori? [Amnesty for plagiarists?]. *G4Media.ro*. https://www.g4media.ro/amnistie-pentru-plagiatori-usr-atrage-atentia-ca-un-amendament-care-i-scoate-basma-curata-pe-plagiatorii-de-doctorate-a-fost-introdus-fraudulos-in-raportul-final-al-comisiei-de-invatamant-din-parl.html.

OECD. (2024). *Education at a glance 2024: OECD indicators*. OECD Publishing. https://doi.org/10.1787/c00cad36-en

Pantazi, R. (2012, June 29). Ministerul Educației a schimbat componenta Consiliului National de Atestare a Diplomelor, institutia abilitata legal sa analizeze acuzațiile de plagiat în cazul tezei de doctorat a premierului Victor Ponta. Decizia Ministerului, luata cu o zi înainte de pronunțarea pe cazul Ponta [The Ministry of Education has changed the composition of the

National Council for Diploma Attestation, the institution legally authorized to analyze accusations of plagiarism in the case of Prime Minister Victor Ponta's doctoral thesis. This decision was made a day before the ruling on the Ponta case]. *Hotnews.ro*. https://hotnews.ro/ministerul-educatiei-a-schimbat-componenta-consiliului-national-de-atestare-a-diplomelor-institutia-abilitata-legal-sa-analizeze-acuzatiile-de-plagiat-in-cazul-tezei-de-doctorat-a-premierului-victor-632976.

Pantazi, R. (2014 Decembre 29). Guvernul declanşează amnistierea plagiatorilor [The government launches amnesty for plagiarists]. *Hotnews.ro*. https://www.hotnews.ro/stiri-esential-18961288-guvernul-declanseaza-amnistierea-plagiatorilor-cum-explica-noul-ministru-educatiei-modificat-legea-pentru-singur-cum-poate-renunta-plagiator-gratis-doctoratul-facut-banii-statului.htm.

Pantazi, R. (2022 July 25). Amnistierea plagiatorilor prin noua lege a educației [Amnesty for plagiarists through the new education law]. Edupedu.ro. https://www.edupedu.ro/breaking-amnistierea-plagiatorilor-prin-noua-lege-a-educatiei-orice-plagiator-va-putea-sa-renunte-gratis-la-doctoratul-facut-pe-banii-statului-daca-legea-este-promovata-premierul-ciuca-protagoni/.

Pantazi, C. (2023, February 25). Analiza celei de-a doua cereri de plată din PNRR, de 3.2 miliarde euro, prelungită cu încă o lună [Analysis of the second payment request from the PNRR, of 3.2 billion euros, extended by another month]. *G4Media.ro*. https://www.g4media.ro/exclusiv-analiza-celei-de-a-doua-cereri-de-plata-din-pnrr-de-3-2-miliarde-euro-prelungita-cu-inca-o-luna-comisia-europeana-are-rezerve-legate-de-protectia-avertizorilor-de-integritate-si-decarboni.html.

Păvălaşc, M. (2021, September 16). Dosar de plagiat. [Plagiarism file]. Europa Liberă România. https://romania.europalibera.org/a/seful-directiei-de-inspectie-constesta-plagiatul/32268647.html.

Peticilă, M. (2020, July 14). Cronologia celui mai mare caz de furt intelectual din România - Plagiatul lui Victor Ponta [Chronology of the most significant case of intellectual theft in Romania - Victor Ponta's plagiarism.]. *Edupedu.ro*. https://www.edupedu.ro/cronologia-celui-mai-mare-caz-de-furt-intelectual-din-romania-plagiatul-lui-victor-ponta/.

Popescu, M. (2024, February 12). Despre Codul-cadru de etică și deontologie universitară. Întrebări, nedumeriri [About the framework-code of university ethics and deontology. Questions, perplexities]. *Contributors*. https://www.contributors.ro/despre-codul-cadru-de-etica-si-deontologie-universitara-intrebari-nedumeriri/.

PressOne. (2024, August 20). Decizie: Mircea Geoană a plagiat ,"43 de rânduri și 2 tabele". Ce nu a luat în considerare Comisia de etică de la ASE [Decision: Mircea Geoană plagiarized „43 lines and 2 tables". What the ASE Ethics Commission did not consider]. https://pressone.ro/decizie-mircea-geoana-a-plagiat-43-de-randuri-si-2-tabele-ce-nu-a-luat-in-considerare-comisia-de-etica-de-la-ase.

Popoveniuc, B. (2022). *Cui* şi *Quid prodest* etica și integritatea academică? [*Cui* and *Quid prodest* academic ethics and integrity?]. Presented at the International Conference *Ethical values in today's society. New challenges and trends (VESA)* organised by Transilvania University in Braşov, in partnership with the University Ethics and Management Council (CEMU) Braşov, 03–05 June 2022.

Puşcaş, F. (2016). Victor Ponta a contestat verdictul cu privire la teza sa de doctorat [Victor Ponta contested the verdict regarding his doctoral thesis]. Stiri pe Surse. https://www.stiripesurse.ro/victor-ponta-a-contestat-verdictul-cu-privire-la-teza-sa-de-doctorat_1143397.html.

Roşca, A. (2022a, July 8). Alin Gavreliuc. Ca să înțelegem ce ne ține pe loc, trebuie să înțelegem această trăsătură: individualismul autarhic. Andreea Roşca. https://www.andreearosca.ro/alin-gavreliuc/.

Roşca, I. (2022b, January 23). Fenomenul plagiatelor. [The phenomenon of plagiarism]. *Hotnews.ro*. https://www.hotnews.ro/stiri-educatie-25320765-fenomenul-plagiatelor-peste-80-000-titluri-doctor-acordate-romania-din-1990-pana-azi-anul-cele-mai-multe-doctorate.htm.

Schiermeier, Q. (2012). Romanian prime minister accused of plagiarism. *Nature*, *486*(7403), Art. 7403. https://doi.org/10.1038/486305a.

Semeniuc, S. (2022, February 16). Cum a mușamalizat de trei ori Comisia de Etică a Universității București analiza acuzațiilor de plagiat din lucrarea de doctorat a judecătoarei Corina Corbu, șefa Înaltei Curți de Casație și Justiție [How the Ethics Commission of the University of Bucharest concealed three times the analysis of plagiarism accusations in the doctoral thesis of judge Corina Corbu, head of the High Court of Cassation and Justice]. *G4Media.ro*. https://www.g4media.ro/exclusiv-cum-a-musamalizat-de-trei-ori-comisia-de-etica-a-universitatii-bucuresti-analiza-acuzatiilor-de-plagiat-din-lucrarea-de-doctorat-a-judecatoarei-corina-corbu-sefa-inaltei-curti-de-casatie-si.html.

Semeniuc, F. (2023). Greșeală elementară sau mână de ajutor? [Elementary mistake or helping hand?]. *G4Media.ro*. https://www.g4media.ro/greseala-elementara-sau-mana-de-ajutor-pentru-ca-a-depus-recursul-in-afara-termenului-legal-ministerul-educatiei-a-pierdut-definitiv-procesul-in-care-premierul-ciuca-obtinuse-suspendarea-analizei-te.html.

Semeniuc, S. (2024, September 19). "O Justiție în slujba ostașului". Serie de evenimente ciudate din dosarele de pe rolul instanțelor care au analizat plagiatul din teza de doctorat a lui Nicolae Ciucă. ["Justice at the service of the soldier". A series of strange events from the files before the courts that analyzed the plagiarism in Nicolae Ciucă's doctoral thesis] G4Media.ro. https://www.g4media.ro/o-justitie-in-slujba-ostasului-serie-de-evenimente-ciudate-din-dosarele-de-pe-rolul-instantelor-care-au-analizat-plagiatul-din-teza-de-doctorat-a-lui-nicolae-ciuca.html.

Semeniuc, F., & Tăpălagă, D. (2022). Cum a fost direcționat dosarul premierului Nicolae Ciucă prin metoda „coperta" către judecătorul Marius Iosif, care s-a pensionat imediat după ce a anulat sesizările de plagiat [How Prime Minister Nicolae Ciucă's file was directed by the "cover" method to judge Marius Iosif, who retired immediately after canceling the plagiarism notifications]. *G4Media.ro*. https://www.g4media.ro/exclusiv-cum-a-fost-directionat-dosarul-premierului-nicolae-ciuca-prin-metoda-coperta-catre-judecatorul-marius-iosif-care-s-a-pensionat-imediat-dupa-ce-a-anulat-sesizarile-de-plagia.html.

Șercan, E. (2023a, May 29). Radiografia unei mușamalizări, frază cu frază (I) [Unraveling a cover-up, phrase by phrase (I).]. *PressOne*. https://pressone.ro/radiografia-unei-musamalizari-fraza-cu-fraza-i-cne-confirma-ca-lucian-bode-a-plagiat-de-cel-putin-26-de-ori-dar-il-declara-neplagiator.

Șercan, E. (2023b, February 22). Urgia la Academia de Poliție (XI) [Wrath at the Police Academy (XI)]. PressOne. https://pressone.ro/urgia-la-academia-de-politie-xi-decizie-istorica-ministerul-educatiei-inchide-fabricile-de-plagiat.

Șercan, E. (2023c, April 27). Un dosar penal și patru acțiuni civile [One criminal case and four civil actions]. PressOne. https://pressone.ro/un-dosar-penal-si-patru-actiuni-civile-cum-incearca-lucian-bode-sa-anuleze-decizia-de-plagiat-emisa-de-universitatea-babes-bolyai.

Șercan, E. (2024a, August 26). Cum a îngropat Comisia de etică a ASE plagiatul lui Mircea Geoană într-una dintre cele mai rușinoase decizii luate vreodată. [How the ASE Ethics Commission buried Mircea Geoană's plagiarism in one of the most shameful decisions ever made] PressOne. https://pressone.ro/exclusiv-cum-a-ingropat-comisia-de-etica-a-ase-plagiatul-lui mircea-geoana-intr-una-dintre-cele-mai-rusinoase-decizii-luate-vreodata.

Șercan, E. (2024b, August 29). Cazul Geoană. Raportul de neplagiat "manipulează brutal cadrul legislativ pe care îl invocă" și ne ține captivi în "noaptea minții". [The Geoana case. The no-plagiarism report 'brutally manipulates the legislative framework it invokes' and keeps us captive in the 'night of the mind'.] *PressOne*. https://pressone.ro/cazul-geoana-raportul-de-neplagiat-manipuleaza-brutal-cadrul-legislativ-pe-care-il-invoca-si-ne-tine-captivi-in-noaptea-mintii.

Tăpălagă, S. & Semeniuc, F. (2024, February 15). Procurorul Iorga Moraru a clasat o anchetă esențială referitoare la plagiatul lui Ciucă [Prosecutor Iorga Moraru closed an essential investigation regarding Ciucă's plagiarism]. *G4Media.ro*. https://www.g4media.ro/exclusiv-procurorul-iorga-moraru-a-clasat-o-ancheta-esentiala-referitoare-la-plagiatul-lui-ciuca-a-ajuns-sefa-sectiei-de-combatere-a-coruptiei-din-dna-la-cateva-zile-dupa-inchiderea-anchetei.html.

Tuomasjukka, T. (2023, October 31). Romania: Political advertising in media is further corroding independent journalism. International Press Institute. https://ipi.media/romania-political-advertising-in-media-is-further-corroding-independent-journalism/.
Voiculescu, L. (2022, October 5). "Dar ce îmi ceri mie să fiu eu super riguros?!...". Efectele plagiatelor asupra studenților de la ASE [But why are you asking me to be super rigorous?!... The effects of plagiarism on ASE students]. *Republica*. https://republica.ro/zdar-ce-imi-ceri-mie-sa-fiu-eu-super-riguros-efectele-plagiatelor-asupra-studentilor-de-la-ase.

Chapter 13
Establishing the Link Between Contract Cheating and Admissions Fraud: A Case Study

Darcey Dahl and Brandon Ng

Abstract The relationship between admissions fraud and contract cheating has been conceptualised as part of a single ecosystem of academic fraud, but there are few case studies on this topic. In this chapter we address the gap in limited research by discussing a case study at the University of New South Wales (UNSW), Australia, that demonstrates how a large-scale investigation established the link between admissions fraud and contract cheating from an investigator's perspective.

The UNSW experience indicates that the connection between contract cheating and admissions fraud is strong. The case study will showcase a procedure-driven, investigative approach to detection of a large cluster of admissions fraud and its relationship to contract cheating. Our investigation, triggered by an allegation of admissions fraud, uncovered a group of 43 international students who were admitted to UNSW between 2017 and 2019 based on fraudulent documentation. Of these, 40 students enrolled at UNSW, with misconduct investigations proceeding against 36 students as 4 had already been permanently excluded due to contract cheating. The investigation resulted in findings of admissions fraud against all 36 students with a substantial proportion having engaged in contract cheating, resulting in permanent exclusion for the students involved.

Recommendations stemming from the investigation include regulatory reforms to prevent academic fraud, incorporating back-to-source checks on admission documents in contract cheating investigations and likewise checks for contract cheating when investigating admissions fraud, and adoption of credentialing platforms for easy verification of qualification records to improve the integrity of the admissions process and protect the reputation of degrees and graduates, as well as protect the wider public.

Keywords Admissions fraud · Contract cheating · Academic fraud detection · Investigative approach · Regulatory reform · Credentialing platform

---

D. Dahl (✉) · B. Ng
University of New South Wales, Sydney, NSW, Australia
e-mail: darcey.dahl@unsw.edu.au; b.ng@unsw.edu.au

M. Davis, C. Baptista (eds.), *Ethics and Integrity in Education (Practice)*, Ethics and Integrity in Educational Contexts 10,
https://doi.org/10.1007/978-3-031-95686-7_13

Introduction

Contract cheating, Paper Mills, Degree Mills and Admissions Fraud underpins what Eaton and Carmichael (2022) describe as "The Ecosystem of Commercial Academic Fraud". However, there are few published case studies that support their proposition as this relationship remains understudied (Eaton & Carmichael, 2023). This chapter will discuss the relationship between the elements of the academic fraud ecosystem from the investigator's perspective at the University of New South Wales (UNSW), Sydney, Australia - namely admissions fraud and contract cheating, as well as the participants and stakeholders in the supply chain. It will also discuss opportunities for institutional and regulatory reforms that can positively disrupt the lifecycle and potentially prevent student misconduct.

At UNSW, the admissions process and the enforcement of academic integrity are independent functions, with different divisional oversight, and rarely intersect on a regular basis. This functional structure has likely impacted the institution's ability to establish a link between contract cheating and admissions fraud. Prior to our investigation, the link between admissions fraud and contract cheating at UNSW had not been established because admissions fraud was, up until mid-2019, investigated by a separate Division and usually occurred at the time of application for admission or within a student's first semester of enrolment.

What we know here at UNSW is that cheating in admission applications is real, is happening at scale, and is a 'hi viz' integrity red flag for our institution, our higher education sector in Australia, and for education providers internationally. Falsifying education credentials is not new; however, digitisation has monetised fake degrees and commercialised, on an industrial scale, fraud in admission applications.

We consider this is also an opportunity for a call to action by all in the academic integrity space, particularly those whose role is to investigate fraudulent student conduct in the higher education sector. Admissions fraud and contract cheating are internationally recognised high risks for any institution. It threatens the integrity of any merit based, equal opportunity admissions process. It also threatens the reputation of our degrees and graduates and more significantly, it can threaten the public health and safety of the local, national, and international communities we serve.

What we also know is that UNSW, and this case study, is not unique. UNSW is unique, however, in the Australian landscape for being open and transparent in what is happening. UNSW is the first and only Australian University to publicly report on its student complaints and conduct in an annual report with the intent that by shining a light on this, sector-wide action can be taken.

Regulatory Landscape: Key Governing Bodies in Australia

In Australia, tertiary education is regulated to ensure the integrity and quality of qualifications, and that students receive education that meets national standards. Two key national regulatory agencies responsible for overseeing tertiary education include the Tertiary Education Quality and Standards Agency (TEQSA) for higher education and the Australian Skills Quality Authority (ASQA) for vocational education and training.

TEQSA is the national regulatory agency for the Australian higher education sector, which "includes public and private universities, Australian branches of overseas universities, university colleges and institutes of higher education" (TEQSA, 2022).

ASQA is the national regulator for vocational education and training in Australia ensuring that registered training organisations (RTO) delivering post-compulsory education and training that provides occupational or work-related knowledge and skills comply with standards set by the Australian government (ASQA, 2015). ASQA accredits qualifications and courses to ensure they meet national standards and needs of industry. ASQA oversees compliance with those standards by conducting regular audits and investigations and has the authority to take regulatory action against RTOs that do not meet national standards.

The Conduct and Integrity Office at UNSW

Who We Are and What We Do

At UNSW, the Conduct and Integrity Office (CIO) supports the UNSW community to uphold honest and ethical behaviour at the university.

> *The UNSW Conduct & Integrity Office ensures that the pursuit of knowledge at UNSW is achieved with respect and integrity. Working collaboratively with students, academics, and professional staff, we resolve complaints and allegations of wrongdoing within a defined framework to cement UNSW as Australia's global university, improving and transforming lives through excellence, research, outstanding education, and commitment to advancing a just society.* (UNSW Conduct and Integrity Office, 2024).

CIO is responsible for complaint management, student misconduct, research misconduct, and other forms of serious wrongdoing, serving as a regulatory body for students at the University. Its mandate spans across a spectrum of responsibilities, from promoting awareness of expected behaviours including academic integrity to investigating allegations of serious misconduct. CIO is committed to fostering an environment of academic excellence and ethical integrity and does this through outreach, education, and enforcement.

CIO is nationally and internationally recognised for innovation in detecting and managing contract cheating (QII, TEQSA, President's Awards) and is endorsed by the UK House of Commons for its Courageous Conversation Program as an innovative and educative approach to student misconduct (Parliamentlive.tv, 2021). We are also the first Australian university to publish transparent annual reports on student conduct, complaints, gendered violence and research integrity (UNSW Sydney, 2022a).

CIO is comprised of several functional areas including: (a) Research Integrity, (b) Disclosures and Complex Investigations, (c) Coordination, Advice and Support, (d) Prevention and Engagement, (e) Systems and Reporting, and (f) Student Conduct and Integrity (SCI) which the investigators of this case study form part of. Each respective area comprises professional staff who are subject matter experts in their fields, bringing a wealth of experience to their roles. Backgrounds and skills are diverse, but all within CIO share a common purpose to support efforts to prevent misconduct and wrongdoing and promote respect and integrity.

SCI investigates student breaches of the *UNSW Code of Conduct & Values*. These breaches can be either academic or non-academic in nature. Examples of academic misconduct include contract cheating, exam misconduct, the submission of false documents, serious plagiarism, and the unauthorised use of Generative AI. Non-academic breaches include behaviour that is discriminatory, unlawful, or inappropriate or damage to university property.

Our Investigative Approach

Before we detail the case study it is useful to understand the typical evidence CIO uses to substantiate cases of contract cheating which can be illustrated as follows:

Figure 13.1 shows the three main categories of evidence that are taken into consideration when the appointed investigating officer and the relevant final decision

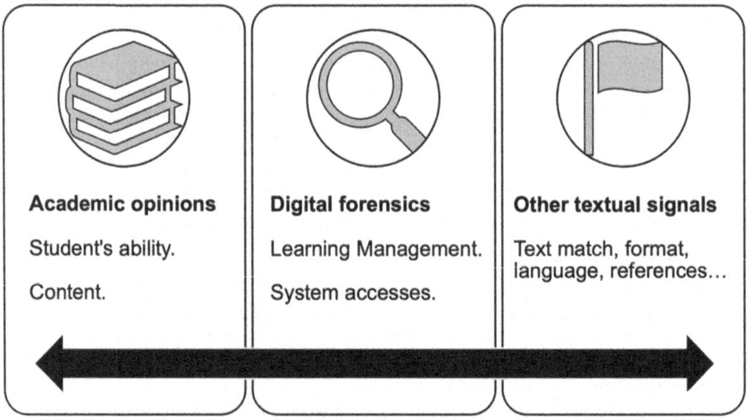

Fig. 13.1 Typical evidence: contract cheating (adapted from Dahl et al., 2023)

maker considers the weight of evidence. These are also in line with TEQSA's recommendations to contract cheating investigators (TEQSA, n.d.).

CIO strictly adheres to making an informed evidence-based decision in contract cheating matters and bases investigations and decision-making processes on these multiple sources of evidence. Often, this includes any combination of the evidence in Figure 13.1, and it is often necessary for investigating officers to proactively collect evidence through data and information that were readily available within the confidentiality and privacy agreement that the student consented to upon admission.

The case study will first discuss how CIO uncovered evidence that a group of students were fraudulently admitted into UNSW and will then examine in detail the academic fraud of one of these students.

Case Study

Admissions Fraud

The role of Registered Training Organisations and education agents.

In March 2021, the UNSW Admissions team detected fraudulent academic documentation submitted by the Sydney office of an international education agent in support of an international student admissions application. The fraudulent academic documents were detected during a back-to-source check after the student was admitted and included an academic transcript and degree certificate purported to be issued by a Sydney based Registered Training Organisation (RTO).[1] Following the detection of the fraudulent documentation, the matter was referred to CIO for further investigation under the UNSW *Student Misconduct Procedure*.

Following a thorough review of the evidence, the student in question was formally contacted by a senior investigator. The student claimed that any issue with their documentation was due to either staffing or system error within the RTO or because of their education agent, who happened to be listed as one of the RTO's Student Recruitment Partners. This was the second student within 9 months who was alleged by the UNSW Admissions team to have submitted fraudulent academic documentation from this RTO. Findings were substantiated against the first student resulting in their permanent exclusion from UNSW. As the investigator considered the probability of such a coincidence to be highly unlikely, the scope of the investigation widened to include any student who submitted academic documentation from the RTO from 2016 onwards as well as the name of the education agent in question who submitted the admissions application on the student's behalf.

[1] A Registered Training Organisation is an Australian education provider that delivers Vocational Education and Training (VET) courses and qualifications. This RTO provided VET courses and qualifications in Business Administration, Leadership and Management, Marketing and Communication, and Project Management.

Further investigation revealed that similar fraudulent academic documentation from the RTO was submitted by the same Sydney based international education agent in support of 43 international student applications dating back to 2016. All 43 students were from China.

After several unsuccessful attempts to confirm the authenticity of the academic documents, the senior investigator arranged a meeting with the RTO's leadership team in June 2021 with the Director, CIO, in attendance. The RTO's Chief Executive Officer was unaware of their organisation's academic documents being used in admissions fraud and suggested it was an issue between UNSW and the education agent, while revealing a personal connection with the education agent's co-founder.

After considerable prompting by the investigator, the RTO provided CIO with confirmation that the documents submitted to UNSW were not genuine. Only 18 of the 43 students had been enrolled at the RTO; however, none of them completed their qualifications. All 43 students were purportedly enrolled in the same course program at the RTO, the Diploma of Business Administration. Of the 43 applications submitted, 40 students accepted the offer of admission and enrolled at UNSW Sydney. Investigations proceeded against 36 students as 4 students had already been permanently excluded from UNSW because of findings of contract cheating.

All the fraudulent academic documents were certified by and submitted through the Sydney branch of a well-known international education agent, who also happened to be one of Australia's top agents in terms of enrolment numbers for onshore international student recruitment. UNSW notified the Chief Operating Officer of the international education agent of the fraudulent academic documentation detected, and suspended all recruitment through their onshore offices pending the outcome of an investigation into the matter. The education agent was asked to provide a report on how the fraudulent documents came to be submitted to UNSW. They were also asked to provide a list of all sub-agents they utilised.

The report was received within a week of the request and a meeting was held with executives from the education agent, UNSW leadership and the lead investigator. The education agent denied any knowledge of the fraudulent documentation from the RTO, but they did find that the documents originated from one of their sub-agents. The relationship with this sub-agent had been terminated in 2018 following a suspicion of fraud, but they had failed to inform UNSW of this at the time. The education agent also confirmed that they would no longer accept academic documents from the RTO involved.

The education agent advised that back-to-source checks had been conducted on some of the academic documents in question, but not all. From internal documents provided to CIO from the education agent, we discovered that the fraudulent documents appear to have been falsely certified by a staff member previously employed by the RTO. This employee was a contact of the sub-agent who had been terminated. While no evidence was found that the education agent had been aware of the fraudulent documents, the meeting highlighted significant deficiencies in their verification processes.

Furthermore, documents provided to UNSW from the education agent revealed that several of these 18 students applied for an early release letter from the RTO after a minimum of 6 months and were granted the requested release, though the standard duration of the Diploma of Business Administration was one-year, full-time study. This meant that the 18 students never completed their qualification at the RTO. Once the students received their early release letter, the sub-agent would then submit the fabricated academic documents to the education agent within 24 h for the purposes of applying for admission to UNSW.

Student Investigations and the Evidence

Once the RTO confirmed the academic documents were not genuine, the student misconduct investigations commenced against each of the students. 13 students engaged with the investigation by replying to correspondence from CIO investigators.

The fraudulent admission documents included academic transcripts and degree certificates with completion dates ranging between 2016 and 2019. Scrutinising the academic documents, the following inconsistencies were identified:

- The Principal Executive Officer's (PEO) signature had been forged as could be seen by the numerous variations in their signature. The PEO later confirmed that the signature appearing on the academic documents was not theirs.
- Study and completion dates were inconsistent with ASQA requirements, specifically that qualifications must be issued within 30 days of a student meeting the requirements of their program.
- Dates were absent on several transcripts.
- The results were not consistently written, that is, some transcripts showed 'C' and others 'Competent'.
- Unit of competency (UoC)[2] enrolment dates on the transcripts are inconsistent. Some showed 3 terms while others showed 4.

Figures 13.2 and 13.3 below are examples of the fraudulent transcripts.

In Australia, a Unique Student Identifier (USI) is a reference number issued by the Australian government to all Vocational Education and Training (VET) students from 2015 and for higher education students from 2023. This number is unique to all students and is used to create an online record of any qualifications and training a student has completed. It also allows students to provide consent to authenticate their qualifications and training through a VET transcript on the official USI website. A VET transcript provides details about a student's completed units of

[2] As vocational qualifications are nationally accredited, every Australian RTO, as part of mandatory compliance, must use the same qualification name and associated code (e.g. BSB50415 Diploma of Business Administration). Units of competency titles and codes are also standardised across the vocational education and training sector. There were no discrepancies identified with qualification and unit of competency names and codes.

Qualification: **BSB50415 Diploma of Business Administration**
A qualification recognised within the Australian Qualifications Framework
Completed on: 3 December 2017

TERM/YEAR	Unit Code	Unit Title	RESULTS
Term 1/2016	BSBADM502	Manage meetings	C
Term 1/2016	BSBITB501	Establish and maintain a workgroup computer network	C
Term 1/2016	BSBADM504	Plan and implement administrative systems	C
Term 2/2016	BSBPMG522	Undertake project work	C
Term 2/2016	BSBMGT517	Manage operational plan	C
Term 2/2016	BSBADM506	Manage business document design and development	C
Term 3/2016	BSBWOR501	Manage personal work priorities and professional development	C
Term 3/2016	BSBSUS501	Develop workplace policy and procedures for sustainability	C

THESE UNITS HAVE BEEN DELIVERED AND ASSESSED IN ENGLISH

Fig. 13.2 Example of a fraudulent transcript featuring an incorrect completion date

Qualification: **BSB50415 Diploma of Business Administration**
A qualification recognised within the Australian Qualifications Framework
Completed on: 19th October 2018

TERM/YEAR	Unit Code	Unit Title	RESULTS
Term3/2017	BSBADM502	Manage meetings	Competent
Term3/2017	BSBITB501	Establish and maintain a workgroup computer network	Competent
Term 4/2017	BSBADM504	Plan and implement administrative systems	Competent
Term 4/2017	BSBPMG522	Undertake project work	Competent
Term 1/2018	BSBMGT517	Manage operational plan	Competent
Term 1/2018	BSBADM506	Manage business document design and development	Competent
Term 2/2018	BSBWOR501	Manage personal work priorities and professional development	Competent
Term 2/2018	BSBSUS501	Develop workplace policy and procedures for sustainability	Competent

Fig. 13.3 Example of a fraudulent transcript displaying irregular results

competency, courses, and qualifications. Without a USI, the RTO could not have legally issued the diploma to these students. Of the 13 students that engaged with our investigations, none were aware of, nor had a registered USI and none could provide a copy of their VET transcript.

Of significant concern was the active involvement of education agents. All but one of the students who responded maintained that any fraudulent issues were the fault of the education agent they engaged to help them gain admission to UNSW Sydney, but no evidence of this contact or agreements have been sighted. One of the sub-agents named was included on the RTO's list of preferred agents.

Of those students who engaged with our investigations, we received evidence showing a pattern of predatory behaviour by their education agents, who facilitated a fraudulent scheme to ensure admission to UNSW. For instance:

- There was no formal contractual arrangement with the education agents and communication and payments were arranged informally via WeChat.
- Payments of up to AU$20,000 were made to the education agents who "promised" admission to UNSW.
- Students paid tuition and fees to the education agent and never directly to the RTO.
- Several students reported that they submitted assessments to their education agent and not directly to the RTO.

Principally, CIO could have concluded its investigations based on the admissions fraud alone where the standard penalty is permanent exclusion. However, considering that 24 of these students had completed more than a year of study, the decision was made to commence investigations into contract cheating to determine those courses the students legitimately passed and could remain on their academic record. The following section will detail the investigation into the academic fraud by one of these students.

How Admissions Fraud and Contract Cheating Can Overlap

To illustrate how close the relationship between admissions fraud and contract cheating can be, this section will focus on a student whose transcript was verified as non-genuine. For this case study, the student will be referred to as "Alex". Alex was a postgraduate student based in Sydney. At the time the investigation started, Alex had already graduated from a UNSW undergraduate program and was currently undertaking the Master of Commerce program with only 3 months of studies remaining.

Our comprehensive investigations of admissions fraud and academic misconduct resulted in the following allegations:

1. Alex submitted fraudulent academic documentation to UNSW in support of their admission application to their undergraduate course.
2. There was inconsistent document metadata in Alex's assessment submissions in over 50 documents across 22 courses taken at UNSW for both their undergraduate and postgraduate studies. This included: 20 different authors and/or "last-modified-by" names in the document metadata; eight different software versions

to create documents; and inconsistent use of page formats between A4 and US letter.

3. Alex's learning management system (LMS) account was accessed from multiple locations, including IP addresses located in Kenya, Russia, Canada, Hong Kong, China, Japan, and the United States. The accesses sometimes occurred simultaneously.

While it is not this chapter's intention to go into detail about how to prove contract cheating (or cybersecurity breaches), it is useful to understand that high inconsistencies in document metadata of assessment submissions and learning management systems being accessed from multiple locations, are strong indicators of contract cheating, and would likely suggest that third parties were accessing LMS to access learning materials, submit assessments on Alex's behalf, and possibly more (such as downloading other confidential data, UNSW and third party copyrighted materials, intellectual property or undertaking other types of cyber fraud).

All UNSW students are bound by the *UNSW Student Code of Conduct*, which outlines the standard of conduct expected of all students. In this case, Alex breached the following specifics of this code (UNSW, 2021):

- Student responsibility 1: Comply with the University's conditions of enrolment by meeting the admission requirements as the evidence suggested that their admission documents were fraudulent.
- Student responsibility 2: Behaving ethically and with integrity, avoiding any action or behaviour that would unfairly advantage or disadvantage either themselves or another student, including contract cheating, fabricating and falsifying documents as the evidence suggested that they submitted falsified documents, and they contract cheated throughout their studies.
- Student responsibility 5: Use and care for University resources in a responsible and appropriate manner, specifically by keeping their login credentials secure and must not disclose to third parties as the evidence also suggested third parties logged into Alex's LMS account.

The Investigation Process

Preliminary Steps

Before the allegations were presented to Alex and the investigation officially commenced, CIO implemented academic transcript and graduation blocks on their record. The purpose of these blocks was to prevent Alex from obtaining a copy of their academic transcript until the investigating officer completes the investigation and the decision maker can decide on the appropriate penalties. This also prevents the student from graduating before the investigation is completed.

This step is crucial, as a copy of academic transcript at that stage would allow Alex to apply to other higher education institutions with a genuine academic transcript, either in Australia or overseas, as any outcome of the investigation would not have been accounted for. Simply put, without this block, Alex could immediately apply to study at another university, perhaps with some credit transfer, and any potential misconduct findings that occurred at UNSW would not matter as it is not currently possible to share misconduct findings with third party institutions and Alex would have legitimised their fraudulent behaviour through another institution.

Communication with the Student

CIO presented the allegations to Alex in the form of an allegation letter in accordance with the *UNSW Student Misconduct Procedure*.[3] This procedure was designed to address alleged breaches of the *UNSW Student Code of Conduct*, and it emphasised *procedural fairness*, ensuring a fair and just process throughout, based on evidence (Fig. 13.4).

Alex responded with the following after a series of correspondence:

1. Alex stated they had never seen any of the fraudulent admission documents. Alex stated that prior to being admitted to UNSW, they knew a person on WeChat, a social media app, who purported to be an education agent with a well-established agency. For the purpose of this case study, we refer to this contact as "Danny". Alex consulted Danny about the likelihood of being admitted into UNSW and provided details of their education background.

   Upon review, Danny advised Alex that their qualifications did not meet UNSW's entry requirements, but for the price of A$15,000 Danny would be able to "repackage" Alex's application to guarantee entry into UNSW.

   Alex paid Danny without seeing or signing any documents and they were eventually accepted into UNSW. Alex stated that while they acted negligently throughout the application process, it was not their intention to submit any fraudulent documentation to the university.
2. Alex denied having engaged in contract cheating.
3. Alex also denied any disclosing of their UNSW login credentials to third parties. They stated that they spent a substantial amount of time away from Australia due to the COVID-19 pandemic and had to study online from mainland China where their family lived. Because of this they had to use various VPN services to log into their LMS to avoid any internet censorships.

[3] This procedure has since been replaced by the *UNSW Complaints Management and Investigations Policy* and Procedure since this investigation took place, following the same spirit and principle as the previous *UNSW Student Misconduct Procedure*.

UNSW takes procedural fairness seriously.

Procedural fairness underpins our investigations and decision making at every step. At UNSW, we aspire to practice procedural fairness by being impartial, even-handed and evidence-based decision makers.

Procedural fairness in our investigative practice means describing which student responsibility of the UNSW Student Code of Conduct was breached, clear framing of the evidence and including any observations such as an academic's opinion, or comments from a witness regarding the conduct. The UNSW Student Misconduct Procedure mandates that when managing an investigation UNSW will:

- Provide the student with an opportunity to respond to the complaint of misconduct and to any questions raised within a stipulated time frame.
- May request to meet with the student. The student may bring a Support Person to the meeting.
- Provide the student with the draft Investigation Report for comment on any omissions or errors of fact.
- If a decision maker determines that a disciplinary outcome is appropriate, they will request a submission from the student on any mitigating circumstances.

When we investigate, using the same rules of evidence as would be applied by a court or tribunal, on any subsequent challenge by the student (e.g. internal appeal or external escalation), it is likely that the factual findings will be similar and will withstand scrutiny by an external third party such as a court or tribunal, or ombudsman.

Fig. 13.4 Procedural fairness at UNSW

The Investigation Discussion and Findings

Having considered Alex's response to the evidence and observations, the Investigating Officer assessed the response alongside the evidence and discussed the following main points with Alex:

1. Irrespective of whether the student used an education agent to assist with the admission process into UNSW, the admission documents are expected to be true and correct in every aspect, whether it was submitted by the student themselves or by an agent on Alex's behalf. Furthermore, Alex failed to provide evidence supporting their engagement with Danny.
2. Whilst Alex denied the contract cheating allegations, they failed to provide a credible explanation to counter the overwhelming evidence of not being the sole author of the submitted assessments.
3. Whilst Alex denied they disclosed their UNSW login credentials to third parties and explained that the accesses from multiple locations were caused using VPNs,

the explanation did not stand as the IP addresses were checked against commercial IP quality checkers and confirmed that they were not VPN related. This meant that they were likely accessed from the person who generated the activity in the location of the IP address.

Given that Alex had admitted to gaining enrolment into UNSW based on fraudulent documents and was unable to provide any evidence to support their denial to contract cheating and disclosing their UNSW login credentials, the Investigating Officer made the following findings based on the *balance of probabilities*:

1. Alex submitted fraudulent documentation for admission into UNSW.
2. Alex submitted assessments in 22 courses that were not all their own work and had been produced with the assistance of third parties.
3. Alex provided their UNSW login credentials to third parties.

The investigation report and its attachments were then sent to Alex to identify any errors or misstatements. As none were identified, the Investigating Officer then forwarded the investigation report and its attachment to a decision maker within CIO, who determined Alex's allegations were substantiated and proposed that Alex receive the following penalties:

– EM (Excluded Misconduct)[4] grade for all 22 courses, spanning across their undergraduate and postgraduate studies.
– Permanent exclusion from UNSW.

Before these penalties were implemented, the decision maker requested the student's submission of any mitigating circumstances or consideration into the penalties. Alex subsequently responded, admitting to having used a "grammar assistance service", apologised for the inconvenience caused and hoped the university can waive the permanent exclusion and accept them as a student to start over again as an undergraduate student.

While the decision maker noted Alex's submission, the overall circumstances did not warrant a lighter penalty under the *UNSW Student Misconduct Procedure*. The proposed penalties were implemented, and it resulted in Alex's undergraduate degree being revoked as they no longer met the degree requirements.

Alex did not appeal the decision.

[4] The definition of EM grade – "Student has been permanently excluded on the grounds of academic misconduct." (UNSW Sydney, 2022b) The grade is applied to each individual course and overwrites any marks/grades the student received for those subjects.

Findings and Institutional Response

Table 13.1 outlines the key outcomes of the investigations. In total, 43 students were admitted to UNSW based on the fraudulent documentation; however, three students did not enrol as they did not accept the offer of admission to UNSW. Four students had already been permanently excluded because of substantiated findings of contract cheating. Thus, investigations proceeded against 36 students. All were found to have submitted fraudulent documents for admission, and findings of contract cheating were substantiated against a further 8 students. It is also notable that 9 students had other prior misconduct findings.

Following our investigation, the disciplinary outcome for the 36 students with a substantiated finding of admissions fraud included permanent exclusion from the University and an EM grade for any course where contract cheating was also substantiated. While these consequences may seem harsh to some, we believe they are appropriate because these students lacked the necessary qualifications for entry to UNSW. In several cases, they had also engaged in other forms of student misconduct. Moreover, we note in New South Wales, using false documents is possibly unlawful behaviour according to section 254 of the *Crimes Act 1900* (NSW).

Only one out of the 36 students that we investigated appealed, which ultimately was not upheld. We consider that this is a remarkably low appeal rate and serves as a testament to the effectiveness and fairness of our investigative process. Furthermore, it suggests that the outcome of these investigations was justified, reflecting robust adherence to procedural integrity and good decision making by the investigating officers and decision makers.

This investigation led to changes in our standard operating procedures. Prior to this cluster of investigations, admissions fraud and contract cheating were investigated as separate breaches of the *UNSW Student Code of Conduct*. Now, given a clearly identified connection between the two credential frauds, if there is alleged contract cheating, we will query the admission documents and if we receive an allegation of admissions fraud, we will run concurrent investigations for contract cheating if coursework has been completed.

In 2022, following the conclusion of the investigations, CIO:

- notified ASQA requesting that they investigate the RTO for non-compliance with the *Standards for Registered Training Organisations 2015* (https://www.asqa.gov.au/about-us/asqa-overview/key-legislation/standards-rtos-2015) under the Vocational Education and Training quality framework.

Table 13.1 Key investigation outcomes

Total number of students	Students who did not accept offer of admission	Students already excluded	Investigated and permanently excluded	Students Completing >1 Year	Students with prior misconduct findings	Contract Cheating Substantiated
43	3	4	**36**	24	9	8

- reported the findings of the investigation to the New South Wales Independent Commission Against Corruption (ICAC) as required under section 11 of the *Independent Commission Against Corruption Act 1988* (New South Wales Independent Commission Against Corruption, 1998). Section 8 of the Act defines corrupt conduct as any action by a person that could influence the execution of official duties by public officials. As UNSW is a publicly funded institution and its staff are public officials, admissions fraud is considered corrupt conduct according to the Act.

The institutional response was swift and led to several actions taken by UNSW:

- All student recruitment through the education agent's onshore offices was suspended pending the findings of their report on how the fraudulent academic documentation came to be submitted to UNSW and evidence that the education agent implemented significant improvement in their internal documentation verification process. Suspension was implemented in compliance with UNSW's obligations under Standard 4 of the National Code.
- CIO widened the scope of their investigation to review the academic documentation of other RTOs which were submitted through the education agent's Sydney office.
- UNSW reviewed the processes in place to ensure adequate oversight of sub-agent activities.

Disrupting the Lifecycle of Admissions Fraud and Contract Cheating

The outcomes of our investigations indicate existing admission processes are imperfect, at best. The sector needs to engage in an open and honest dialogue about admissions fraud and how it could call into question a student's approach to academic integrity. A student's willingness to commit admissions fraud could suggest a propensity towards dishonest behaviour that could make them more likely to engage in contract cheating once admitted. We want to see all universities identify fraud - in all its forms - as a ubiquitous severe risk and managed appropriately. To ensure both students and the public have trust and confidence in the admission process (and ultimately, graduates), universities and governments must work together to identify and address systemic causes of admissions fraud and contract cheating.

Our investigations indicate education agents were integral to the credential fraud ecosystem (Eaton & Carmichael, 2022). While fabricated academic documents and contract cheating, on their own, are not a new phenomenon, the monetisation and digitisation of international student recruitment has compromised admission procedures and exposed a vulnerability to predatory agents. We consider information sharing among universities about RTOs, other institutions and education agents implicated in admissions fraud as a vital step in the disruption process.

Institutional Reforms

To disrupt the lifecycle of admissions fraud and contract cheating, there must be reform within universities to make fraud free admission an institutional priority. As a first step, institutions must commit to and invest in an appropriately resourced admissions office and dedicated team to detect admissions fraud. We want to have confidence in university processes and know that those students admitted are in fact genuine applicants who have met entrance requirements.

Admissions fraud is a complex issue with competing tensions. Two of those obvious tensions are the financial imperative to enrol students and maintaining the integrity of our admission process. Both are equally important, and one should not trump the other. There are inherent conflicts of interest that exist between the international education agents, universities, and the students and these need to be mitigated. There is a constant pressure on universities, such as reduced government funding (Smith, 2020), to maintain enrolment numbers and the revenue stream, and many Australian universities have international student enrolments as a main key performance indicator.

A back-to-source check is the most commonly used way to verify an applicant's credentials; however, in its current form we consider that it is the weakest way to capture fraud as the University often does not get the documents verified, whether because of sheer volume, the issuing institution does not respond, or another insurmountable bureaucratic impasse. To combat this, we recommend all institutions globally to come to an agreement on the best way to facilitate back-to-source checks. This may include the use of a regulated and independent credentialing platform for verifying tertiary academic documents similar to "My eQuals" Australia (My eQuals, 2024) or the use of a secure centrally managed depository, like the USI system in Australia, can be advocated to consolidate training and qualification records.

While we acknowledge administrative barriers, universities should, where possible, facilitate information sharing. This may include:

- Facilitating requests from other institutions for back-to-source checks.
- Establishing student misconduct registers for students who were found to have systematically engaged in contract cheating.
- Establishing registers of dishonest education agents.

The potential link between admissions fraud and contract cheating should spur university staff and administrators at all levels to act. For example, front line university staff should consider, at a minimum, incorporating back-to-source checks on students where they are found to have systematically engaged in contract cheating to assure the quality of students we are graduating.

It is also important to have confidence in our admission procedures. Institutions should conduct annual independent audits of admission and education agent processes. Institutions should also consider mandating the use of authorised education agents who have undergone extensive scrutiny in high-risk areas to mitigate exploitation from individuals purporting to be education agents.

Sector Wide and Regulatory Reforms

Universities cannot tackle this problem alone. Governments must take steps to reform the sector. Standard 4 of the National Code of Practice for Providers of Education and Training to Overseas Students, states that all Australian universities are required to ensure their agents act ethically, honestly and in the best interests of overseas students (Australian Government, Department of Education, 2023). Universities are also required to monitor agents to ensure that they are compliant with their agency agreement in accordance with this standard (Australian Government, Department of Education, 2023). The onus is on universities to ensure the education agents act ethically; however, state and federal governments can do more.

We believe the federal government should establish a national register of international education agents in Australia with the following functions:

- Set national standards and policies that all registered international education agents must meet.
- Ensure agents are suitably trained, qualified and accredited. This should expand to the sub-agents as well.
- Apply consistent standards in recruitment delivery.
- Publish an online register of agents terminated by education providers for conduct in breach of their agency agreement, or other unethical conduct.

Some individuals claiming to be education agents are sometimes paid by students with a promise to get them admitted into their chosen university and program. Authorised education agents of UNSW are paid commission by the university for each eligible student who is admitted (and remains enrolled) as per their agent agreement. Several of the students who engaged with our investigations were purportedly defrauded by the education agent, having paid them as much as A$20k as a separate fee to any admissions or visa processing fees. Because of this, we recommend consumer protections for duped international students.

Limitations

Despite the institutional response resulting from our investigation, several limitations should be acknowledged:

Limited Resourcing

The prevalence of contract cheating and admissions fraud is likely underreported, and universities do not know the true extent of the problem (Curtis et al., 2021). Detection is difficult and substantiation of fraud even more so at an institutional level. Investigations require significant resources, including experienced staff and access to detection tools, and this can be costly for any university.

With current resourcing, it is unrealistic to complete a back-to-source check on every application, considering the UNSW Admissions team processes between 15,000 and 20,000 admissions applications each year.

Lack of Cross-Institutional Cooperation

There is a lack of institutional appetite to share data with other institutions and admit there is a problem with the admissions process. Stricter controls at one institution, without similar controls at others, could influence competitiveness in international student recruitment leading to a potential impact on revenue streams if enrolment numbers are affected.

Privacy laws in applicable jurisdictions, such as the Australian Privacy Act 1988 and General Data Protection Regulation (GDPR) in the European Union, may prevent the cross-institutional sharing of information. This could include contracts the universities hold with education agents and student data.

Even if an institution excludes a student after identifying admissions fraud or contract cheating, the student can easily apply for admission to another institution who would be unaware of the misconduct. Likewise, if an institution suspends or terminates an agent, the fraudulent applications can be rerouted to other institutions who would be none the wiser.

Roadblocks to Conducting Back-to-Source Checks

Many institutions especially those located in non-English speaking countries fail to respond to a back-to-source request making it very difficult to verify academic documents. An increasing number of institutions require payment for verification services during the back-to-source process.

While the use of an independent credentialing platform for verifying academic documents would be a step in the right direction, who would regulate the independent credentialing platform and how would it be funded?

Future Directions

While our student misconduct investigations established a link between contract cheating and admissions fraud and brought attention to the need for greater regulation of education agents, the purpose of this chapter is to describe our experience as investigators and inform university administrators and regulators. Therefore, we acknowledge that this chapter describes a single case study. In future, formal research studies are required to establish the extent to which admissions fraud is associated with contract cheating and the prevalence of these forms of misconduct in various student cohorts. Secondly, the fact that 9 of the 36 students investigated in this case study had prior student misconduct findings against them raises the possibility that admissions fraud may be associated with other kinds of misconduct. Future research could also examine whether admissions fraud is specifically associated with contract cheating or whether it generalises to various types of student misconduct.

Conclusion

This chapter has discussed the relationship between admissions fraud and contract cheating and how it overlaps in the ecosystem as framed by Eaton and Carmichael (2022). Our investigations have underscored the need for comprehensive strategies to combat contract cheating and admissions fraud including institutional commitment, regulation of education agents, adequate resourcing to verify admissions documentation and increased investment in detection of contract cheating. While there is more work to be done, this case study suggests that students who engaged fraudulently at the admission stage, are also likely to engage in contract cheating and possibly other types of student misconduct. We believe that if we can stop the admission of students with non-genuine credentials, we can potentially prevent contract cheating.

Acknowledgements The authors wish to extend their gratitude to Julia Lines for her steadfast and invaluable support, as well as her insightful contributions to this work. Special thanks to Bron Greene, Matt Sherwood, and Shaun Khoo for their valuable reviews and strong support of this chapter. We also extend our gratitude to all our hard working colleagues at the Conduct and Integrity Office at UNSW.

References

Australian Government, Department of Education. (2023, September 19). *National Code of Practice for Providers of Education and Training to Overseas Students 2018*. https://www.education.gov.au/esos-framework/national-code-practice-providers-education-and-training-overseas-students-2018.

ASQA. (2015). Asqa.gov.au. https://www.asqa.gov.au/resources/glossary/v.

Curtis, G. J., McNeill, M., Slade, C., Tremayne, K., Harper, R., Rundle, K., & Greenaway, R. (2021). Moving beyond self-reports to estimate the prevalence of commercial contract cheating: An Australian study. *Studies in Higher Education, 47*(9), 1844–1856. https://doi.org/10.1080/03075079.2021.1972093

Dahl, D., Lines, J., & Ng, B. (2023, July 12–14). *The supply chain of contract cheating: the sum of its parts – a case study of investigating admission fraud and contract cheating at the University of New South Wales (UNSW)* [Conference Presentation]. 9th European Conference on Ethics and Integrity in Academia (ECEIA), Derby, UK.

Eaton, S. E., & Carmichael, J. (2022). The ecosystem of commercial academic fraud. *Werklund School of Education Research & Publications.* Calgary, Canada: University of Calgary. http://hdl.handle.net/1880/115324.

Eaton, S. E., & Carmichael, J. J. (2023). Fake degrees and credential fraud, contract cheating, and paper mills: Overview and historical perspectives. *Fake Degrees and Fraudulent Credentials in Higher Education, 5*, 1–22. https://doi.org/10.1007/978-3-031-21796-8_1

My eQuals. (2024). *Home*. https://www.myequals.edu.au/.

New South Wales Independent Commission Against Corruption. (1998). *Independent Commission Against Corruption Act 1988*. https://legislation.nsw.gov.au/view/whole/html/inforce/current/act-1988-035.

Parliamentlive.tv. (2021, February 10). *Ten minute rule motion: Essay mills (Prohibition)* [Video]. https://parliamentlive.tv/event/index/1145a875-5853-49d1-b606-a7abb5e83272?in=13:59:10.

Smith, L. (2020, May 25). *How universities came to rely on international students*. The University of Sydney. https://www.sydney.edu.au/news-opinion/news/2020/05/25/how-universities-came-to-rely-on-international-students.html.

TEQSA. (n.d.). *Substantiating contract cheating: A guide for investigators*. https://www.teqsa.gov.au/sites/default/files/2022-10/substantiating-contract-cheating-guide-investigators.pdf.

TEQSA. (2022). *Our role | Tertiary Education Quality and Standards Agency*. Teqsa.gov.au. https://www.teqsa.gov.au/about-us/teqsa-overview/our-role.

UNSW Sydney. (2021). *Student code of conduct (v4.0)*. University of New South Wales. https://www.unsw.edu.au/content/dam/pdfs/governance/policy/2022-01-policies/studentcode-policy.pdf.

UNSW Sydney. (2022a). *UNSW leads the way by publishing Student Conduct and Complaints Annual Report*. Inside UNSW. https://www.inside.unsw.edu.au/campus-life/unsw-leads-the-way-publishing-student-conduct-and-complaints-annual-report.

UNSW Sydney. (2022b). *UNSW grade definitions*. https://www.student.unsw.edu.au/grade.

UNSW Conduct and Integrity Office. (2024). *Our purpose*. Unpublished internal document.

Chapter 14
The Face of Fake Science: Four Episodes for Reflection

Artem Artyukhov

Abstract This chapter presents a comprehensive exploration of predatory scientific practices in Ukraine, focusing on the tactics employed by those who prey on unsuspecting scientists. Through a combination of descriptive analysis and personal experience, the author aims to expose the exploitative nature of these practices and caution scientists against falling victim. The chapter begins by highlighting the key characteristics of predatory conferences, such as excessive fees, vague submission guidelines, and a lack of peer review. The author then shares their personal experience of publishing a meaningless article in a predatory journal, demonstrating the ease with which such publications can be achieved.

To provide further insights, the chapter also presents a "portrait" of the "criminal" who preys on scientists, outlining their common tactics and motivations. Additionally, a hypothetical dialogue between a scientist and the forces of fake science illustrates the persuasive techniques used to lure scientists into participating in predatory activities. By combining descriptive analysis with personal experience, this chapter offers a practical guide for scientists to identify and avoid predatory practices. It emphasises the importance of critical thinking, due diligence, and ethical considerations in navigating the complex landscape of scientific research and publication.

Keywords Fake science · Academic integrity · Predatory journal · Pointless article

In a world of facts and dreams,
Where knowledge flows in endless streams,
A shadow lurks, a whispered lie,
In the guise of truth, it flits and flies.

A. Artyukhov (✉)
Sumy State University, Sumy, Ukraine

University of Economics in Bratislava, Bratislava, Slovakia
e-mail: a.artyukhov@pohnp.sumdu.edu.ua

M. Davis, C. Baptista (eds.), *Ethics and Integrity in Education (Practice)*, Ethics and Integrity in Educational Contexts 10,
https://doi.org/10.1007/978-3-031-95686-7_14

With glittering charts and flashy claims,
It dances lightly, plays its games,
A miracle cure, a wondrous find,
But reason falters, and logic's blind.

Oh, the siren call of easy answers,
In the hearts of many, it plants its cancers,
With promises bright, it lures the lost,
But wisdom pays the heaviest cost.

"Believe in me," the charlatan cries,
As it spins its tales beneath the skies,
A potion here, a crystal there,
But science whispers, "Beware, beware."

For every fact that stands so tall,
A pseudo-science seeks to make it fall,
With cherry-picked data and biased views,
It sows confusion, it spreads the blues.

Perplexity (https://www.perplexity.ai/). on the query "write me a poem about fake science"

Introduction

The phenomenon of fake science refers to the spread of false or misleading information in the scientific community, which can lead to misinformation and disinformation (Lazer et al., 2018). This issue has been exacerbated by the rise of social media and the ease with which information can be disseminated globally (Hopf et al., 2019). Fake science can be found in various fields, including archaeology, where fake news existed 3000 years ago (Conley, 2022). The scientific community has recognised the problem of fake science, with the percentage of scientific articles retracted due to fraud increasing significantly since 2000. The incentives for authors and publishers to game the system for high status and profits can lead to the fabrication of data and the creation of predatory journals. The consequences of fake science can be severe, leading to a loss of respect for the results of science and the scientific method, which can result in a steep decline in funding, jobs, and students wishing to enter the field (Hopf et al., 2019).

The academic and scholarly publishing structure means that any fake science operation faces challenges in expanding beyond individual (self-made) sources. However, the potential for fake science to be used offensively, like fake news, cannot be ignored. The more a country invests in its scientific research efforts, the lower the likelihood that the country will try to "poison the well" by spreading fake science (Fletcher, 2019).

To address the problem of fake science, individual scientists, professional associations, academic institutions, and funding bodies must act to promote ethics and integrity. Researchers, educators, and policymakers must contribute to research that helps understand and counter false information and strengthen science literacy in society. The battle for truth requires a collective effort to put the house of science in order and counter false information in circulation (Hopf et al., 2019).

The Ukrainian scientific community is periodically rocked by scandals such as "Theory of Non-Force (Information) Interaction" by Iurii Teslia (Filtenborg & Weichert, 2019) and "A Lepton 'God'" by Kateryna Kyrylenko (Sharapov, 2016), and does not always properly react to fake science, because the history of fake science has deep roots and is not something new in the cultural space of Slavic countries. "Vernalization of winter crops and genetics" by Trofim Lysenko (Borinskaya et al., 2019), "the process of transformation of inanimate matter into living" by Olga Lepeshinskaya (Rejón, 2018), "japhetidology" by Niko Marr (Allen, 2020) are far from a complete list of "sensational" theories.

A significant part of bad practices still present in the Ukrainian scientific community have their roots in the historical background. Fake science can have a devastating effect on public institutions and security at all levels - from the health and education of citizens to the development of state programmes and strategies (Stemwedel, 2011). As a result, there is a need to come up with measures to wipe out the toxic legacy of the Soviet Union in terms of fake science still practised in the Ukrainian community. Fake science can be commissioned or used to achieve formal scientific heights (Doctor of Sciences, Academician). At the same time, fake ("predatory") journals and conferences are created to promote pseudo-scientific results; for the sake of results, scientists engage in several types of academic dishonesty.

The presented work covers four episodes, four authors' observations on the nature of fake science. In the first episode ("*Portrait*"), the author tries to make a portrait of a "criminal" who hunts for ignorant (although sometimes ideally understanding that they are doing wrong) scientists. In the second episode ("*Dialogue*"), the author's fantasy presents an imaginary dialogue between a scientist and fake science, trying to drag him to the "dark" side. The third episode ("*Features of the predatory conference, Ukrainian experience*") presents the main signs of "predatory" conferences, the organisers engaged in making money and not holding high-quality scientific events. The fourth episode ("*Pointless article*") summarises the essence of pseudo-scientific publications based on the author's experience publishing a meaningless article in a "predatory" journal. In this episode, there are lessons for scientists who, for the sake of dubious fame, are ready to publish their shoddy scientific materials. The first two episodes are descriptive and show the author's vision. The third and fourth episodes are the author's experiment in participating in a "predatory" conference and publishing in a "predatory" journal. The presented material and description of the episodes will be helpful for both young researchers and experienced scientists. The former will receive information about the dangers in the future scientific path; the latter will receive information about modern methods of deceiving scientists who sincerely want to receive wide recognition through their publications.

Literature Review

Analysis of literary sources on the "Fake Science" topic reveals a growing concern regarding the proliferation of misinformation and pseudo-scientific claims in academic and popular discourse. Various scholarly works and articles have delved into the phenomenon, shedding light on its origins, impacts, and potential remedies. For example, "Merchants of Doubt" by Naomi Oreskes and Erik M. Conway (2011) examines the tactics employed by vested interests to sow doubt and discredit established scientific consensus, particularly in fields such as climate change and tobacco regulation. The authors reveal how these strategies have led to the spread of fake science, undermining public trust in scientific findings and impeding effective policy action. Similarly, "The Death of Expertise" by Tom Nichols (2017) explores the erosion of expertise and the rise of anti-intellectualism in contemporary society. Nichols argues that the democratisation of information, facilitated by the internet and social media, has paradoxically led to a devaluation of expertise and a proliferation of fake science. He highlights the dangers of this trend, emphasising the importance of deference to qualified experts in shaping public opinion and policy.

In addition to scholarly analyses, journalistic investigations have also contributed to our understanding of fake science. Books such as "Bad Science" by Ben Goldacre (2009) and "Science Fictions" by Stuart Ritchie (2020) expose how flawed research practices, biased reporting, and commercial interests contribute to the dissemination of unreliable scientific information. These works underscore the need for critical thinking and rigorous scrutiny of scientific claims to distinguish genuine research from fake science.

Two more books should also be noted:

1. "Denying to the Grave: Why We Ignore the Facts That Will Save Us" by Sara E. Gorman and Jack M. Gorman (2016). This book explores the psychological factors contributing to accepting fake science and misinformation. It delves into cognitive biases, belief perseverance, and the societal implications of scientific denialism.
2. "The Emperor's New Drugs: Exploding the Antidepressant Myth" by Irving Kirsch (2011). Kirsch critically examines the evidence behind the efficacy of antidepressant medications and highlights the influence of pharmaceutical companies on research outcomes. This book sheds light on the issue of publication bias and the manipulation of scientific data.

Fanelli (2009) conducted a systematic review of the literature on scientific misconduct, focusing on fabrication, falsification, and plagiarism. The study found that different institutions adopted different definitions of scientific misconduct, but all agreed that fabrication, falsification, and plagiarism were severe forms of misconduct. Plagiarism was qualitatively different from the other two because it did not distort scientific knowledge, although it had significant consequences for the careers of the people involved. The study emphasised the importance of limiting the dissemination of fake news to mitigate its damaging effects.

Patra and coauthors (2023) conducted a bibliometric analysis of literature published on "fake news" and associated terms such as misinformation, disinformation, and post-truth in various disciplines. The study found that 2014–2020 was the most productive period in studying fake news, with prominent themes in the political, health, technology, media, and social media realms. The study concluded that mapping scholarly contributions of scientific research provides a guiding approach and helps counter the information chaos stimulated by fake news phenomena in the digital era.

Dhiman and coauthors (2023) conducted a scientometric analysis of academic articles on fake news and related subjects, including infodemic, misinformation, and disinformation in pandemics. The study identified various research publishing categories for fake news detection, including linguistic style, implementation of these styles in future work, publication trends based on the year, regions or nations that contributed more to the topic, authors who contributed significantly to the detection of fake news, publishing trends based on connections (colleges/organisations), and citation counts for each of the contributions related to fake news detection methods and techniques.

The study by De Bruyckere (2023) highlights the prevalence of fake scientific papers in neuroscience and medicine, with up to 34% and 24% of papers published in 2020 being, respectively, likely made up or plagiarised. The research emphasises the threat of paper mills, which churn out bogus manuscripts containing text, data, and images partly or wholly plagiarised or fabricated, often massaged by ghostwriters. The recent advent of artificial intelligence tools such as ChatGPT has amplified the concern, with publishers and researchers working together to develop new tools to detect and combat the spread of fake papers.

Fake science poses a significant threat to academic integrity, creating a complex landscape with pros and cons. On the one hand, the rise of fake science undermines the credibility of scientific research and erodes trust in the scientific community. The proliferation of falsified results and fraudulent publications can lead to a loss of respect for the scientific method, impacting funding, job opportunities, and the willingness of students to pursue careers in science. However, the exposure of fake science can also catalyse the improvement of research integrity and the promotion of ethical conduct within the scientific community. By addressing false information and actively combating fake science, researchers, professional associations, academic institutions, and funding bodies can strengthen the foundation of scientific knowledge and uphold the principles of integrity and ethics. In the context of academic integrity, the presence of fake science highlights the importance of promoting transparency, accountability, and rigorous peer review processes. While fake science can be damaging, it also underscores the need for vigilance and critical evaluation of research findings to ensure the reliability and validity of scientific knowledge.

Public dishonesty and unethicality, in their general sense, have many manifestations: In business it comes mainly through the shadow economy (Bilan et al., 2019, 2020a, 2020b; Tiutiunyk et al., 2022) or non-transparent and biased management (Kobiyh et al., 2024; Morin, 2024; Steenbergen et al., 2023), various types of discrimination in the career building (Adams-Harmon, 2024). In society, it happens due to corruption (Bozhenko et al., 2023a), lack of transparency of public institutions (Bozhenko et al., 2023b; Zámek & Zakharkina, 2024), or distortion of information (Užík et al., 2024). In medicine, this results from non-transparency of public reporting (Kolomiiets et al., 2023) or limited patient control (Waldman, 2024). In education it happens due to academic dishonesty, deformation in organisational culture (Brainard, 2023; Kettaf et al., 2024), neglect of the quality of education (Samiya & Asma, 2024), or preferences for certain categories of students (Hara, 2023). In science, it comes through using untested or unconfirmed research methods (Tchoketch-Kebir & Madouri, 2024), and gender bias (Bhowmik, 2023).

Despite the large number of regulatory and advisory documents at the national and international levels, the actual situation in Ukraine is far from ideal. Ukrainian scientists must provide indicators for the contract and reporting indicators for the implementation of grant projects. At the same time, individual scientists or groups of scientists' resort to the help of so-called "scientific consulting" firms, which offer intermediary services for publishing low-quality scientific results in predatory journals and conferences. Another "assistance" option is selling co-authorship or ghost writing. Naturally, the low-quality "products" that needed to be promoted had the properties of "Bad Science" and "Fake Science". This bridge is the subject of study in this work (Table 14.1). This figure highlights the gap between the existence of academic integrity frameworks and the reality of academic misconduct. Despite numerous well-established rules and regulations at international, national, and institutional levels, problems such as fake science, predatory publishing, and ghost writing still result in questionable scientific work. The figure underscores the discrepancy between theoretical regulatory frameworks and the practical challenges of enforcing integrity in academic research and publishing.

Table 14.1 The reality of the Ukrainian scientific community: state of art

<table>
<tr><th rowspan="2">Rules and regulations</th><th colspan="2">Reality</th></tr>
<tr><th>Process (cause)</th><th>Result (effect)</th></tr>
<tr><td>International Center for Academic Integrity</td><td rowspan="3">Fake Science</td><td rowspan="4">"Theory of Non-Force (Information) Interaction" by Iurii Teslia</td></tr>
<tr><td>European Network for Academic Integrity</td></tr>
<tr><td>Council of Europe Platform on Ethics, Transparency and Integrity in Education</td></tr>
<tr><td>The European Code of Conduct for Research Integrity</td><td rowspan="3">Predatory conferences and journals</td></tr>
<tr><td>The Bucharest Declaration on ethical values and principles of higher education in the Europe region</td><td rowspan="5">"A Lepton 'God'" by Kateryna Kyrylenko</td></tr>
<tr><td>Task Force for future monitoring of values. Report to the Bologna Follow-Up Group, Helsinki</td></tr>
<tr><td>National regulatory framework</td><td rowspan="3">Ghost writing firms</td></tr>
<tr><td>Ukrainian scientist Code of ethics</td></tr>
<tr><td>University's Code of Conduct</td></tr>
</table>

Bibliometric Analysis

A bibliometric analysis was conducted to find a connection between the term "Fake Science (Pseudo-science, False Science)" and various aspects of scientific activity. An array of publications was examined, the dataset of which was generated from the Scopus database[1] using the query "Fake Science (Pseudo-science, False Science)." To conduct a bibliometric analysis, the following restrictions were introduced into the search query:

1. Indexation period – 2022-5.04.2024.
2. Branch of knowledge – All areas.
3. Based on this dataset, keyword maps (Figures 2–6, please see Appendix 1 in Artyukhov, 2022) were built. Bibliometric analysis tools VOSviewer[2] and SciVal[3] were used.

General Overview of Figure 3 Pseudoscience is at the centre of the network and is the largest node, suggesting it is the primary concept linking various terms in the dataset. From here, other terms radiate outwards, demonstrating their relevance or frequency of co-occurrence in texts or research related to pseudoscience. The terms around pseudoscience are grouped into clusters of different colours, each representing a different thematic area related to pseudoscience.

Breakdown of Clusters by Colour

Green Cluster (Education and Knowledge)
Includes terms such as "education," "teaching," "students," "knowledge," and "scientific literacy." These terms suggest that one thematic focus of the network is how pseudoscience interacts with educational contexts, perhaps discussing how pseudoscientific beliefs arise in students or how critical thinking and scientific literacy are essential in combating pseudoscience.

Red Cluster (Science and Public Opinion)
Contains terms like "ethics," "public opinion," "politics," "economics," "prevention and control," and "scientists." This cluster seems to deal with societal and ethical dimensions of pseudoscience, including its interaction with politics, public opinion, and economics, as well as how pseudoscience can be controlled or prevented through ethical scientific practices.

[1] https://www.scopus.com/, accessed on April 5, 2024.

[2] https://www.vosviewer.com/.

[3] https://www.scival.com/.

Yellow Cluster (Critical Thinking and Scepticism)
Terms like "critical thinking," "scepticism," "science denialism," and "nature of science" are part of this cluster. This fact likely represents the focus on critical thinking and scepticism as tools to counter pseudoscience. The presence of terms like "science denialism" indicates that the cluster might be addressing the broader problem of rejecting scientific consensus in favour of pseudoscientific ideas.

Purple and Blue Clusters (Methodology and Evidence)
Terms such as "evidence-based," "methodology," "scientific procedure," and "forecasting" suggest a focus on the importance of rigorous scientific methods and evidence-based practices. These terms suggest discussions around scientific methodologies and the need for evidence-based practices to combat pseudoscientific ideas or claims.

Key Concepts

"Critical thinking" and "scepticism" are central to addressing pseudoscience. These terms highlight the educational need to promote a questioning mindset as a defence against false claims.

"Politics" and "public opinion" suggest that pseudoscience is not just a scientific issue but also a social and political one, where misinformation can affect policy making and public beliefs.

"Ethics" and "Scientists": These terms point to the ethical responsibility of the scientific community to address pseudoscience and uphold scientific integrity.

"Education" as a recurring theme indicates the importance of improving scientific literacy and teaching methods to help individuals discern between science and pseudoscience.

Cluster "evidence-based practice" description (Figures 4–6).

Methodology and Evidence-Based Practice: pseudoscience often lacks rigorous methodology and relies on anecdotal evidence or personal beliefs rather than empirical data. This contrasts with the scientific method, which emphasises evidence-based practice and critical thinking.

Education: education plays a crucial role in promoting critical thinking and scepticism, which are essential tools for distinguishing between science and pseudoscience.

Ethics and Public Opinion: pseudoscience can raise ethical concerns, as it may exploit public beliefs and emotions. Public opinion can also influence the acceptance or rejection of pseudoscientific ideas.

Government and Politics: pseudoscience can sometimes be used for political or economic gain, influencing government policies and public discourse.

Technology and Information: the spread of pseudoscience can be facilitated by technology, such as social media and online platforms. Misinformation and disinformation can thrive in these environments.

Marketing and consumer highlight the commercial aspects of pseudoscience, which can be used to promote products or services.

The results of bibliometric analysis also made it possible to establish the following patterns:

1. Among the most popular topics is fake science on the path of knowledge and technology transfer. This fact points to the importance of increasing scientists' awareness of the negative consequences of fake science.
2. Cluster analysis identified nine clusters:

   - prevention, control, politics;
   - emotion, motivation, decision making;
   - the core of fake science;
   - prediction and education;
   - media and communication;
   - ethics and open science;
   - scientific knowledge;
   - culture;
   - awareness.

At least five out of nine clusters are related to the need to promote the principles of real science and consider cases of fake science.

3. Fake science is considered, from the point of view of ethics and public opinion, an essential tool for promoting verified science.
4. Quite unexpected is the presence of a cluster that links fake science, economic indicators and government measures to combat fake science. This fact indicates the need for local popularisation events and mandatory consideration of cases (Problem-based learning and Evidence-based learning).
5. AI is seen as one of the tools for the possible emergence of fake science and attention is paid to violations of academic integrity.
6. Fake science is viewed through the prism of conspiracy theory. It is necessary to increase scientists' awareness of the methods of fact-checking and verification (falsification) of information.
7. The clusters highlight the concepts of "awareness" and "responsibility", which suggests that fake science is an object of unfavourable influence on the image of a scientist.

The literature review and bibliometric analysis results prompted the author to present four episodes (cases) of the promotion of verified science, described below.

Episodes

In this section, the author tries to present an algorithm for introducing a scientist with fake science. This scientist is full of faith that all scientific research can be verified or falsified based on its results (facts). This scientist has tools (arguments, judgments, results of previous research) to verify or falsify research. However, the first encounter with fake science differs from the introduction to science. Therefore, the author offers cases that are created based on analytical materials (for example, Skeptical Science (2021) or Luoma (2006) for an easier understanding of the essence of fake science.

Episode 1: Features of the Predatory Conference, Ukrainian Experience
This episode delves into the characteristics of predatory conferences, drawing on the author's experiences in Ukraine. It highlights the red flags that scientists should be aware of, such as excessive registration fees, vague submission guidelines, and a lack of peer review. The episode aims to educate researchers on how to identify and avoid these deceptive events.

Episode 2: Pointless Article
In this episode, the author shares their personal experience of publishing a meaningless article in a predatory journal. By highlighting the ease with which such publications can be achieved, the episode serves as a cautionary tale for scientists who may be tempted to prioritise quantity over quality in their research. It emphasises the importance of maintaining high standards and avoiding the pursuit of dubious fame.

Episode 3: Portrait
This episode paints a vivid picture of the "criminals" who operate predatory conferences and journals. It explores their motivations, tactics, and the techniques they use to deceive unsuspecting scientists. By understanding these individuals, researchers can better protect themselves from falling victim to their schemes.

Episode 4: Dialogue
In this episode, the author imagines a hypothetical dialogue between a scientist and the forces of fake science. The dialogue illustrates the persuasive techniques used by predatory entities to lure scientists into participating in low-quality events or publishing in questionable journals. By understanding these tactics, scientists can be better prepared to resist such attempts and make informed decisions about their research.

Episode 1: Features of the Predatory Conference
(Ukrainian experience, based on Ukrainian Kitchen for Scientific Publications (n.d.), moderator - Serhiy Kozmenko, and own investigations).

- remote (not online) Internet conference;
- "scientific and practicality";
- foreign international conference - the languages of publications are Ukrainian and Russian;
- proceedings submission for 7 days before the publication of "articles";
- fake bibliographic information: ISBN and publisher;
- a broad scope of the conference;
- acceptance of abstracts "at the last minute" (even on the conference day);
- payment is expected immediately;
- payment accounts related to an individual instead of the organisation;
- spam warning.

Episode 2: Pointless Article

The author's creative approach in describing the four episodes is not an entertaining tool but an attempt to popularly explain the challenges to scientists and the dangers of "predatory" publications. The author presents his experiment of publishing in a "predatory" journal (Artyukhov, 2022) (Figure 7, please see Appendix 1).

In this article, the author writes (original text from an article with intentional grammatical errors): "Before the reader is an example of compiling and using a set of terms to describe processes that are not related to each other. In the conditions of today's rapid development of the technical field and in view of the imperfection of local systems for working with scientometric data, in particular, the motivation of scientific employees for the regularity of original research of a certain quality, researchers are forced to resort to partial fabrication of results. Such an approach cannot be called useful for the field in the full sense, but it results in a certain improvement of scientific texts and the preservation of valuable scientific ideas within closed scientific communities. The submitted material calls for the fight against dishonesty by refusing to publish in dubious journals, participating in "predatory" conferences, and improving qualifications for a certificate in exchange for a certificate. The example of the material presented here and the fact that it was accepted for publication shows that scientists have the opportunity (often for a moderate price) to publish pseudo-scientific articles and, as the heads of various organisations of the "Worldwide Development of Scientific Research" like to say (the example of the title is the author's), to expand the "geography" of publicising one's achievements and to give significance to one's profession through internships and publications abroad."

The original correspondence (screenshots) with the main stages of article submission history is placed in Appendix 2.

A reproduction of this correspondence is shown in Table 14.2.

Episode 3: Portrait

Where to find it?

- officially: nowhere;
- informally: penetrates the mind at the slightest provocation;
- physically: on the pages of predatory journals and conferences;
- mentally: hidden in each of us.

General omens:

- promises a simple path to success and worldwide recognition;
- does not recognise alternative opinions and relies on faith;
- offers emotions instead of facts and evidence;
- evaluates science only qualitatively;
- positions itself as an opponent of traditional science;
- looking for a conspiracy against itself.

Special omens:

- says what you want to hear;

Table 14.2 Article submission history

Sender	Stage	Day	Mail content (original grammar and punctuation)
Publisher	Invitation from the publisher	Day 0. Start of counting	Dear Dr. Artyukhov Artem Hope you are doing well We are in need of articles for successful release of Volume 4 Issue 3 by 3lst October? Is it possible for you to support us with your 2 page opinion/case report or mini review, we hope 2 page article Isn't time taken for eminent people like you We hope you won't disappoint us Acknowledge this email within 24 h
Artem Artyukhov	My response	Day 3	Dear Colleague thank you so much for the invitation. It's an honour for me Please find my paper in the attachment
Publisher	Information about article acceptance of the consideration and request for APC payment	Day 5	Dear Dr. Artyukhov Artem Thank you for submitting your manuscript After having glance at your manuscript we let you know that, your article has been effectively accepted for galley proof and will get back to you with the article updates timely Track the below manuscript ID for further communication Title: Problems of Observing Academic Integrity: The Influence of the Polymer Science Current State Manuscript ID: PSPRJ-22-OP-631 Moreover, we request you to fill the details of your cover letter and send back to us Before proceeding further process of your manuscript, we would like to bring your notice that as AM MS is an open access journal there would be a Nominal Publication Fee towards publishing your article in our Journal Hope you understand us and await your rapid comeback Have a Good day!!
Artem Artyukhov	My response	Day 5	Dear colleagues, thank you for the excellent news! I didn't quite understand the Information about the Nominal Publication Fee. Will you pay me royalties?

(continued)

Table 14.2 (continued)

Sender	Stage	Day	Mail content (original grammar and punctuation)
Publisher	Publisher's response with the clarification of APC amount	Day 5	Dear Dr. Artyukhov Artem Thank you for your reply Well, let me inform that as PSPRJ is an open access journal there will be nominal charges for the publication of article. As we don't get any funding from external organisations and institutions, completely depends upon authors kind contribution only. We request authors to pay minimum publication fee based on their research grants and funds Moreover, we will charge $749 for the publication of your article in our journal. We request you to fill the details in the cover letter and send back to us Hope you understand our concern and await your response Please feel free to contact us for further queries
Publisher	Publisher's information about article acceptance and the requirement for APC	Day 7	Dear Dr. Artyukhov Artem Hope this mail finds you in high spirits Well, let me inform that your article entitled "Problems of Observing Academic Integrity: The Influence of the Polymer Science Current State" is successfully accepted for galley proof. We have forwarded your manuscript to our web department and will get back soon with article PDF Meanwhile, we request you to proceed for the payment $749 for the publication of your article in our journal. Please track the below link for online payment Hope you understand our concern and await your prompt response Have a nice day!

(continued)

Table 14.2 (continued)

Sender	Stage	Day	Mail content (original grammar and punctuation)
Artem Artyukhov	My deliberate silence without response.		
Publisher	Publisher's letters with a reminder of APC	Days 11, 12, 13 and 13 from other email	Dear Dr. Artyukhov Artem Hope this mail finds you in high spirits May I kindly know your acknowledgement regarding the previous mail? Hope you understand our concern and await your prompt response Have a nice day!
			Dear Dr. Artyukhov Artem Hope this mail finds you in high spirits May I kindly know your acknowledgement regarding the previous mail? Hope you understand our concern and await your prompt response Have a nice day!
			Dear Dr. Artyukhov Artem Hope this email finds you in good spirits! We are glad to inform you that we had done with PDF of your article. Please find attached PDF and let us know if any further modifications that you are willing us to do. Moreover, we request you to proceed for the payment $749 for the publication of your article in our journal. Please track the below link for online payment Hope this is fine for you Await your rapid response
			Dear Dr., Artem Artyukhov This is *(Name)*. Hope you are well and having a good time Tomorrow is Friday and we have an audit tomorrow, so we must show the article which got published online. Hence, we are friendly requesting you to take care of the article fee online using the same PayPal link, here you do appreciate it if you try and pay now If you need bank details or any alternate link, sure we can provide you please let me know Title: Problems of Observing Academic Integrity: The Influence of the Polymer Science Current State Journal: Polymer Science: Peer Review Journal Article ID number; PSPRJ-22-OP-631 The publication fee is only $749 but to be honest if you want to publish the same paper in other top journals like us they will charge you more than $1500 but here it is very cheap I hope you will respond to this email at your earliest convenience If you have any questions, feel free to reply to me (address) Trust me we are holding our breath to publish your article online as soon as you make the payment Please respond. Thank you very much

(continued)

Table 14.2 (continued)

Sender	Stage	Day	Mail content (original grammar and punctuation)
Artem Artyukhov	My response	Day 16	Dear *(Name)* thank you for your mail If I'm not mistaken, the editorial board contacted me regarding the article's writing (I wasn't the initiator), in this case, if the editorial board asks to write an article for the journal, it must pay the Honorarium to me. How much money can I expect? Thank you
Publisher	Publisher's response with APC reduction	Days 19 and 20	Hope you are safe and healthy Let me inform you this is Polymer Science: Peer Review Journal Title: Problems of Observing Academic Integrity: The Influence of the Polymer Science Current State Here is the PayPal link *(link)* Art ID: PSPRJ-22-OP-631 and publication fee $500 after waiver before it was $749 This Is my final request sir so please think twice before you reply Await for your answer
			Hi, Dr. Artem Artyukhov Hope you are staying safe and sound I'm happy to inform you that your opinion paper was already published online. Please check the pdf link below (link). If you need more corrections, please let me know so we can do it now I can republish online permanently Could you please take care of the payment now if you do that, we will be thankful to you. Publication fee $500 after waiver before it was $749 We keep an eye to see your positive response to complete the publication process with the activation of the DOI number
Artem Artyukhov	My response	Day 20	Please see Appendix 2
Publisher	Publisher's response with APC requirement. There was no reaction to my previous email	Day 21	Hello, Dear, Artem Artyukhov It's good to know information One more we are cheerful to inform you that we have already published the article online Here is the link to locate the article (link). If you need any help please do reply as we need on waiting for your article to publish Title: "Problems of Observing Academic Integrity The Influence of the Polymer Science Current State" Journal: Polymer Science Peer Review Journal Could you please take care of the payment today because tomorrow is Friday and we are much how the publish articles online I hope you will understand our concern Here is the PayPal (PayPal number) Art ID: PSPRJ-22-OP-631 and publication fee $500 after waiver before it was S749 We hope that you will keep up your word by paying the fee

- always finds like-minded people;
- provide a strong esprit de corps.

Operation: an offer one cannot refuse.

Tactics: gradual "intake" into one's arms.Strategy: no facts, just belief.

Strengths:

- there are many gullible people among scientists;
- for the sake of fame, some are ready for anything;
- "scientists" are willing to pay if only their articles were published.

Weaknesses:

- revelation comes fast;
- money does not solve everything;
- image loss is irreparable;
- it is difficult to return to "traditional" science;
- many scientists know about fake science;
- "bubble" of fake science bursts from a light "factual" touch.

What is the main lesson from studying the portrait of fake science?

Pseudo-science often offers simple solutions to complex problems, tapping into individuals' desires for quick answers and immediate gratification. This emotional appeal can be particularly potent in times of uncertainty.

When the first acquaintance with fake science occurs "in absentia," the scientist is drawn into the funnel of selling success indicators (KPI for a contract or other necessary indicators). The time comes to communicate with a fake scientist or a fake science service provider.

Episode 4: Dialogue

Fake science (FS): Want to be successful?Me (M): What can you suggest?FS: Follow me.M: Why?FS: Trust me.M: Reason?FS: I always tell the truth.M: What is truth?FS: The truth is what you want to hear.M: What about facts?FS: Facts… Mmm… I'll give you positive emotions and greatness in the scientific world.M: Where is the evidence?FS: Wholeheartedly believe me.M: No, I need to think.FS: Why do you need it? I'll lead you to success. No complicated steps.M: No, I need to check the facts.FS: Um… Perhaps you'll believe it? Cause it's a conspiracy theory against me.M: Only facts. Nothing more.FS: But the facts do not please you.M: Truth is only a list of facts. For one person, this list is suitable, for another person - no.FS: Are you leaving me?M: I'd never be with you.

After engaging with fake science, a scientist must weigh the pros and cons of engaging with fake science. The last two episodes will help them in making a decision, which is no longer a fantasy (a well-founded fantasy, but you cannot talk face to face with fake science...) but indicate ways to lure a scientist into the world of simplified quantitative indicators of the scientist's productivity (you will not get quality indicators here). Some may see tempting outcomes in these episodes, while others may see the danger of losing their image.

These four episodes demonstrate the dangers a scientist may face when looking for a place to publish scientific results. The results themselves may be good, but the place of publication will "spoil" them, and this will blow the scientist's image and ambitions.

1. Episode 1
   - Link to Figure 1: this episode connects directly to the "Predatory conference and journals" box under Processes (Cause) on the right side of the figure. The episode describes the red flags of predatory conferences, such as high fees, vague guidelines, and lack of peer review, mirroring the deceptive practices that undermine academic integrity, despite the existence of numerous rules and regulations on the left side of the figure.
   - Connection to reality: Figure 1 emphasises the presence of predatory conferences that, while violating ethical standards, still flourish, leading to the creation of dubious scientific content, as reflected in the examples of works by Iurii Teslia and Kateryna Kyrylenko.
2. Episode 2
   - Link to Figure 1: this episode aligns with both "Fake Science" and "Predatory conference and journals" in the Process (Cause) section. Publishing a meaningless article in a predatory journal exemplifies how fake science and low-quality publications proliferate, often driven by the pressure to publish without concern for quality.
   - Connection to reality: the article's publication in a predatory journal contributes to the production of untrustworthy scientific literature, which the figure illustrates as the Result (Effect) in the form of questionable works like those of Iurii Teslia or Kateryna Kyrylenko. This episode serves as an example of how such "pointless" publications emerge.
3. Episode 3
   - Link to Figure 1: this episode ties to the "Ghost writing firms" and "Predatory conference and journals" sections under Processes (Cause). The "criminals" operating these predatory platforms exploit the academic system by using deceptive tactics, mirroring the unethical behaviour discussed in the figure. This episode paints a picture of those responsible for creating platforms that facilitate fake science.
   - Connection to Reality: the existence of these "criminals" results in a proliferation of fraudulent work, just as the figure highlights how such predatory actors contribute to the publication of dubious scientific results.
4. Episode 4
   - Link to Figure 1: this episode reflects the persuasive nature of predatory entities, aligning with "Fake Science", "Predatory conference and journals", and "Ghost writing firms" in the figure. The dialogue represents the manipulation and tactics used by such predatory actors to lure scientists into unethical prac-

tices, including publishing in questionable journals or attending low-quality conferences.
- Connection to reality: the dialogue illustrates how fake science is constructed and perpetuated, reinforcing the gap between the strict rules (on the left side of the figure) and the actual fraudulent practices (on the right). This dynamic leads to outcomes such as the publication of works like "Theory of Non-Force (Information) Interaction" by Iurii Teslia.

Conclusions

This work attempts to provide a non-trivial description of the face of fake science and the consequences of "entering" the cohort of pseudoscientists. The analysis of Figure 1 and the four episodes are connected by the theme of academic misconduct and the tension between formal rules and real-world challenges in upholding academic integrity. This is how each episode aligns with the processes and effects shown in the figure:

In carrying out scientific activities, a scientist inevitably encounters cases of fake science and organisations that are intermediaries in its promotion. Four episodes on fake science show the evolution in understanding the nature of fake science. This evolution can be described in the following stages:

1. Case studies (episodes 1 and 2): consciously choosing the path of a scientist.

   After stage 3, the scientist (even with a positive attitude towards fake science at stages 1 and 2) can favour science, considering actual image losses. However, as shown, for example, by a preliminary analysis of the services of some providers of "predatory" conferences (analysis of this phenomenon in Ukraine is the subject of subsequent research), participation in this "type" of conferences is widespread.
2. Online introduction to fake science (episode 3): first conclusions about attractiveness.
3. Offline acquaintance with representatives or providers of fake science (episode 3): doubts about the effectiveness.

These four episodes provide valuable insights and practical guidance for scientists and the broader academic community. Here's a breakdown of their practical significance:

1. "Portrait":

   - Awareness: this episode helps scientists recognize the tactics used by predatory entities, enabling them to be more vigilant and avoid falling victim to their schemes.
   - Identification: by understanding the characteristics of these "criminals," scientists can more easily identify potential predatory conferences or journals.

2. "Dialogue":

   - Understanding persuasive techniques: this episode illustrates the persuasive tactics used by predatory entities, helping scientists recognize and resist these tactics.
   - Preparation: by understanding the potential pitfalls of engaging with predatory entities, scientists can be better prepared to navigate such encounters.

3. "Features of the Predatory Conference":

   - Red flag identification: this episode outlines the key characteristics of predatory conferences, allowing scientists to identify red flags and avoid such events.
   - Informed decision-making: by understanding the motivations and practices of predatory conferences, scientists can make more informed decisions about where to submit their work and attend events.

4. "Pointless Article":

   - Cautionary tale: this episode serves as a cautionary tale, highlighting the potential consequences of publishing low-quality work in predatory journals.
   - Maintaining standards: it reinforces the importance of maintaining high standards in scientific research and publication, avoiding the temptation to compromise quality for quick gains.

Call to Action

The following recommendations are suggested for the Higher Education community to stand up to fake science:

1. Be critical and sceptical: approach scientific information with a critical eye. Question claims that seem too good to be true or lack supporting evidence.
2. Verify sources: always check the credibility of sources. Look for peer-reviewed research, reputable publications, and established experts in the field.
3. Educate yourself: stay informed about scientific advancements and developments. Follow reputable science news outlets and resources.
4. Support quality research: support institutions and organisations that promote rigorous scientific research and ethical practices.
5. Expose fake science: if you encounter fake science or predatory practices, do not hesitate to report it to relevant authorities or organisations.
6. Promote scientific literacy: encourage others to develop critical thinking skills and a healthy scepticism towards pseudoscientific claims.
7. Engage in dialogue: participate in discussions and debates about scientific issues, promoting evidence-based reasoning and challenging misinformation.

A Collaborative Approach

1. Building partnerships: collaborating with other scientists, educators, policymakers, and media outlets to raise awareness of the issue and develop strategies to address it.
2. Advocating for policy changes: supporting policies that promote scientific literacy, support rigorous research, and hold accountable those who spread misinformation.
3. Supporting scientific organisations: supporting organisations that work to promote scientific integrity and combat misinformation.
4. Engaging with the public: communicating scientific information effectively and engaging with the public to dispel misconceptions and promote evidence-based thinking.

Appendices

Appendix 1

Figures 2–7 for the chapter "The Face of Fake Science: four episodes for reflection." https://drive.google.com/file/d/1fBy7BheNPBBz6I6UzYMOTIviHhVVrF0L/view?usp=sharing.

Appendix 2

The original correspondence (screenshots) with the main stages of article submission history for the chapter "The Face of Fake Science: four episodes for reflection." https://drive.google.com/file/d/147QOfdlNnArA-D8y9ZE2Od6N-ULMMOyK/view?usp=sharing.

References

Adams-Harmon, D. (2024). The intersectionality experiences of executive, female minorities ascending in STEM: A qualitative, transcendental, phenomenological study. *SocioEconomic Challenges, 8*(1), 1–15. https://doi.org/10.61093/sec.8(1).1-15.2024

Allen, M. C. (2020). Japhetic grammatology. *Interventions, 24*(5), 742–795. https://doi.org/10.1080/1369801x.2020.1784029

Artyukhov, A. (2022). Problems of observing academic integrity: The influence of the polymer science current state. *Polymer Science, 4*(3). https://doi.org/10.31031/psprj.2022.04.000588

Bhowmik, D. (2023). Gender inequality in higher education and research. *Business Ethics and Leadership, 7*(3), 108–119. https://doi.org/10.61093/bel.7(3).108-119.2023

Bilan, Y., Srovnalíková, P., Streimikis, J., Lyeonov, S., Tiutiunyk, I., & Humenna, Y. (2020a). From shadow economy to lower carbon intensity: Theory and evidence. *International Journal of Global Environmental Issues, 19*(1–3), 196–216. https://doi.org/10.1504/IJGENVI.2020.114874

Bilan, Y., Tiutiunyk, I., Lyeonov, S., & Vasylieva, T. (2020b). Shadow economy and economic development: A panel cointegration and causality analysis. *International Journal of Economic Policy in Emerging Economies, 13*(2), 173–193. https://doi.org/10.1504/IJEPEE.2020.107929

Bilan, Y., Vasylieva, T., Lyeonov, S., & Tiutiunyk, I. (2019). Shadow economy and its impact on demand at the investment market of the country. *Entrepreneurial Business and Economics Review, 7*(2), 27–43. https://doi.org/10.15678/EBER.2019.070202

Borinskaya, S. A., Ermolaev, A. I., & Kolchinsky, E. I. (2019). Lysenkoism against genetics: The meeting of the Lenin all-union academy of agricultural sciences of August 1948, its background, causes, and Aftermath. *Genetics, 212*(1), 1–12. https://doi.org/10.1534/genetics.118.301413

Bozhenko, A., Krawczyk, D., Hałuszko, K., & Ozarenko, V. (2023a). Data-mining modeling of corruption perception patterns based on association rules. *Business Ethics and Leadership, 7*(4), 181–189. https://doi.org/10.61093/bel.7(4).181-189.2023

Bozhenko, V., Buriak, A., Bozhenko, A., & Roienko, O. (2023b). Transparency and corruption prevention in financing climate action. *Financial Markets, Institutions and Risks, 7*(2), 88–94. https://doi.org/10.21272/fmir.7(2).88-94.2023

Brainard, J. (2023, May 9). *Fake scientific papers are alarmingly common*. Science. https://www.science.org/content/article/fake-scientific-papers-are-alarmingly-common.

Conley, N. (2022, June 6). *Real phenomena that people used to think were fake*. Grunge. https://www.grunge.com/162519/real-phenomena-that-people-used-to-think-were-fake/.

De Bruyckere, P. (2023, May 12). *"Up to 34% of neuroscience papers and 24% of medical papers published in 2020 were probably fake."* From Experience to Meaning. https://theeconomyofmeaning.com/2023/05/12/up-to-34-of-neuroscience-papers-and-24-of-medical-papers-published-in-2020-were-probably-fake/.

Dhiman, P., Kaur, A., Iwendi, C., & Mohan, S. K. (2023). A scientometric analysis of deep learning approaches for detecting fake news. *Electronics, 12*(4), 948. https://doi.org/10.3390/electronics12040948

Fanelli, D. (2009). How many scientists fabricate and falsify research? A systematic review and meta-analysis of survey data. *PLoS One, 4*(5), e5738. https://doi.org/10.1371/journal.pone.0005738

Filtenborg, E., & Weichert, S. (2019, October 29). *Academics in Ukraine fighting against rampant misconduct*. The Scientist. https://www.the-scientist.com/news-opinion/academics-in-ukraine-fighting-against-rampant-misconduct-66644.

Fletcher, A. (2019, August 5). *Is the Threat of 'Fake Science' Real?* Default. https://www.lawfaremedia.org/article/threat-fake-science-real.

Goldacre, B. (2009). *Bad science*. HarperPerennial.

Gorman, S. E., & Gorman, J. M. (2016). *Denying to the grave: Why we ignore the facts that will save us*. Oxford University Press.

Hara, M. (2023). Educational reform for middle-income trap under digitalization: Culprits, challenges, and strategies in the Philippines. *SocioEconomic Challenges, 7*(3), 200–218. https://doi.org/10.61093/sec.7(3).200-218.2023

Hopf, H., Krief, A., Mehta, G., & Matlin, S. A. (2019). Fake science and the knowledge crisis: ignorance can be fatal. *Royal Society Open Science, 6*(5), 190161. https://doi.org/10.1098/rsos.190161

Kettaf, R., Karima, K., & Zohra, D. (2024). Investigating the impact of organizational climate on organizational silence in higher education institutions. *SocioEconomic Challenges, 8*(1), 170–182. https://doi.org/10.61093/sec.8(1).170-182.2024

Kirsch, I. (2011). *The emperor's new drugs: Exploding the antidepressant myth*. Basic Books.

Kobiyh, M., El Amri, A., Sahib Eddine, A., & Oulfarsi, S. (2024). Literature review and theoretical generalizations of the ethics role in business and management: Family business as a case study. *Business Ethics and Leadership, 8*(1), 93–106. https://doi.org/10.61093/bel.8(1).93-106.2024

Kolomiiets, S., Buriak, A., & Fritsak, M. (2023). Transparency of social-economic determinants in developing the efficient public healthcare system. *Health Economics and Management Review, 4*(2), 61–73. https://doi.org/10.21272/hem.2023.2-06

Lazer, D., Baum, M., Benkler, Y., Berinsky, A. J., Greenhill, K. M., Menczer, F., Metzger, M. J., Nyhan, B., Pennycook, G., Rothschild, D., Schudson, M., Sloman, S. A., Sunstein, C. R., Thorson, E., Watts, D. J., & Zittrain, J. (2018). The science of fake news. *Science, 359*(6380), 1094–1096. https://doi.org/10.1126/science.aao2998

Luoma, J. (2006, April 2). *The seven warning signs of bogus science.* Association for Contextual Behavioral Science (ACBS). https://contextualscience.org/blog/the_seven_warning_signs_of_bogus_science.

Morin, S. L. (2024). Name Bias: Application Through the 4-Step Process Using Foundations of Critical Theory in Organizational Settings. *Business Ethics and Leadership, 8*(1), 163–171. https://doi.org/10.61093/bel.8(1).163-171.2024

Nichols, T. M. (2017). *The death of expertise: The campaign against established knowledge and why it matters.* Oxford University Press.

Oreskes, N., & Conway, E. M. (2011). *Merchants of doubt: How a handful of scientists obscured the truth on issues from tobacco smoke to global warming.* Bloomsbury Publishing.

Patra, R. K., Pandey, N., & Sudarsan, D. (2023). Bibliometric analysis of fake news indexed in Web of Science and Scopus (2001–2020). *Global Knowledge, Memory and Communication, 72*(6/7), 628–647. https://doi.org/10.1108/gkmc-11-2021-0177

Rejón, M. R. (2018, December 21). *Beyond lysenko: the Biological experiments of the Russian revolution.* OpenMind. https://www.bbvaopenmind.com/en/science/scientific-insights/beyond-lysenko-the-biological-experiments-of-the-russian-revolution/.

Ritchie, S. (2020). *Science fictions: How fraud, bias, negligence, and hype undermine the search for truth.* Metropolitan Books.

Samiya, B., & Asma, A. (2024). Multidimensional factors to measure quality in education institutions. *SocioEconomic Challenges, 8*(1), 143–153. https://doi.org/10.61093/sec.8(1).143-153.2024

Sharapov, S. (2016, August 12). *A Lepton "God": Proof of the collapse in the Ukrainian scientific degree awarding system.* Vox Ukraine. https://voxukraine.org/en/argumentum-ad-absurdum-en/.

Skeptical Science. (2021, August 10). *11 Characteristics of pseudo-science.* https://skepticalscience.com/11-characteristics-of-pseudoscience.html.

Steenbergen, E. F., Jansen, W. S., & Ellemers, N. (2023). How executive boards set the stage for unethical behavior in the financial sector. *Business Ethics and Leadership, 7*(4), 108–127. https://doi.org/10.61093/bel.7(4).108-127.2023

Stemwedel, J. D. (2011, October 4). *Drawing the line between science and pseudo-science.* Scientific American. https://blogs.scientificamerican.com/doing-good-science/drawing-the-line-between-science-and-pseudo-science/.

Tchoketch-Kebir, H., & Madouri, A. (2024). Research leadership and high standards in economic forecasting: Neural network models compared with Etalon ARIMA models. *Business Ethics and Leadership, 8*(1), 220–233. https://doi.org/10.61093/bel.8(1).220-233.2024

Tiutiunyk, I. V., Zolkover, A. O., Lyeonov, S. V., & Ryabushka, L. B. (2022). The impact of economic shadowing on social development: challenges for macroeconomic stability. *Naukovyi Visnyk Natsionalnoho Hirnychoho Universytetu, 1*, 183–191. https://doi.org/10.33271/nvngu/2022-1/183

Ukrainian Kitchen for Scientific Publications. (n.d.). https://www.facebook.com/groups/449029836041923.

Užík, J., Oleksich, Z., & Dinits, R. (2024). Model of information dissemination in the context of reputation formation of an auditing company: Official sources or "Word of Mouth"? *Financial Markets, Institutions and Risks, 8*(1), 108–122. https://doi.org/10.61093/fmir.8(1).108-122.2024

Waldman, D. (2024). Replace government healthcare with patient-controlled health care. *Health Economics and Management Review, 5*(1), 80–89. https://doi.org/10.61093/hem.2024.1-06

Zámek, D., & Zakharkina, Z. (2024). Research trends in the impact of digitization and transparency on national security: Bibliometric analysis. *Financial Markets, Institutions and Risks, 8*(1), 173–188. https://doi.org/10.61093/fmir.8(1).173-188.2024

The manufacturer's authorised representative in the EU is Springer Nature Customer Service Centre GmbH, Europaplatz 3, 69115 Heidelberg, Germany. If you have any concerns regarding our products, please contact ProductSafety@springernature.com

Printed and bound by CPI Group (UK) Ltd, Croydon, CR0 4YY
30/09/2026
02225588-0003